# ESSENTIALS
# OF
# BIOSTATISTICS

ROBERT C. ELSTON, M.A., Ph.D.

Professor and Head
Department of Biometry and Genetics
Louisiana State University Medical Center
New Orleans, Louisiana

WILLIAM D. JOHNSON, M.S.

Associate Professor
Department of Biometry and Genetics
Louisiana State University Medical Center
New Orleans, Louisiana

## ESSENTIALS OF MEDICAL EDUCATION SERIES

DAVID T. LOWENTHAL, M.D., Ph.D.  /  Editor-in-Chief

Professor of Geriatric Medicine and Adult Development
Professor of Medicine and Pharmacology
Mt. Sinai School of Medicine
New York, New York

F. A. DAVIS COMPANY • Philadelphia

**Library of Congress Cataloging-in-Publication Data**

Elston, Robert C., 1932–
    Essentials of biostatistics.

    (Essentials of medical education series)
    Includes bibliographies and index.
    1. Biometry.    2. Medical statistics.    I. Johnson, William Davis, 1941–    .  II. Title.   III. Series.
    [DNLM: 1. Biometry. WA 950 E49e]
    QH323.5.E47 1987      574'.072      86-24009
    ISBN 0-8036-3122-7

ESSENTIALS OF MEDICAL EDUCATION SERIES

# Editor's Commentary

## To Our Students

This series has been developed to help you review essential biomedical facts and concepts and reinforce basic clinical skills. Because medical science is forever evolving and expanding, you should expect to spend a lifetime building upon a knowledge base gained through your required curriculum. It is our hope that these texts, together with your educational program and faculty support, will help you develop patterns of independent learning and skills of analytical thinking and clinical decision making. Therefore, in many of the volumes, our authors further aim to challenge you with questions and problems similar to those you will experience in classroom and clinical situations as well as in national examinations.

Although a large component of your education will be devoted to scientific theory and technologic principles, we believe that you must not lose sight of the importance of your patient as an individual. Accordingly, we have tried to maintain a humanistic focus where applicable in both the basic science and the more clinical texts in this series.

In this volume, Professors Elston and Johnson provide a clearly written, easy-to-understand, comprehensive discussion of biostatistics. The book has been especially designed both for those who have little or no statistical background and for those who need to review basic concepts. The authors provide realistic examples to clarify the statistical principles described but avoid undue detail and technicalities. Thus, this text will assist both students and practitioners to better understand and interpret scientific literature.

DAVID T. LOWENTHAL, M.D., Ph.D.
EDITOR-IN-CHIEF

v

# Preface

"Biostatistics, far from being an unrelated mathematical science, is a discipline essential to modern medicine—a pillar in its edifice" [Journal of the American Medical Association (1966) 195:1145]. Today, even more so than twenty years ago, anyone who wishes to read the biomedical literature intelligently needs to understand the essentials of biostatistics. It is our hope that this book will provide such an understanding to those who have little or no statistical background and who need to keep abreast of new medical findings.

Unlike most other elementary books on biostatistics, the main focus of this book is not so much on teaching how to perform some of the simpler statistical procedures that may be necessary for a research paper, but rather on explaining basic concepts. Many of the simpler statistical procedures are in fact described, but computational details are included in the main body of the text only if they help clarify the underlying principles. For those who want them, more detailed computational formulae are given in Appendix 1. Similarly, we have relegated to Appendix 1 other details that, if included in the body of the text, would tend to make it difficult for the reader to see the forest for the trees. If you do not care for all the details, simply ignore Appendix 1. If you wish to have the details, read the notes in Appendix 1 concurrently, chapter by chapter, with the rest of the book.

This book has been written at an elementary mathematical level and requires no more than high school mathematics to understand. Nevertheless, you may find Chapters 4 and 5 a little difficult at first. These chapters on probability and distributions are basic building blocks for subsequent concepts, however, and you should study them carefully. The basic concepts of estimation and hypothesis testing are covered by the end of Chapter 7, and the next four chapters cover special statistical methods that are widely used in biological and medical research. In the last chapter we have tried to review the most important concepts introduced in earlier chapters as they relate to a critical reading of reports published in the biomedical literature.

We have attempted to illustrate the statistical methods described with enough examples to clarify the principles involved, but without their being so many and so detailed that the reader is caught up in irrelevant and unnecessary technicalities. We have tried to make these examples realistic and yet easy to grasp for someone with a background in the biological sciences. Many of our examples come from human genetics. Apart from providing ideal examples in the application of probability, genetics is a discipline that underlies all biology, and so should be of interest to the readers of our book. Detailed knowledge of genetics is not, however, necessary to understand the examples.

Many persons have helped in the preparation of this book and it would be impossible to mention them all by name. Special thanks, however, go to Dr. Robert M. Rippey, Dr. Martin Kotler, Dr. James H. Godbold, Dr. Beth Dawson-Saunders, and numerous medical students at LSU who read a draft of the manuscript and offered many suggestions for correction and improvement, and to Ms. Gina M. Drake and Ms. Elizabeth A. Gallmann who painstakingly typed the manuscript. Any errors that remain, however, are solely our responsibility.

Robert C. Elston, M.A., Ph.D.
William D. Johnson, M.S.

# Contents

## CHAPTER 1

# Introduction: The Role and Relevance of Statistics in Medicine

**Key Concepts:** the scientific method; deductive and inductive reasoning; scientific method; statistic, statistics.

## WHAT IS BIOSTATISTICS?

When taking a clinical history, conducting a physical examination, or requesting laboratory analyses, radiographic evaluations, or other tests, a physician is collecting information (data) to help decide diagnostic and therapeutic actions. The decisions reached are based on knowledge obtained during training, from the literature, from experience, or from some similar source. General principles are applied to the specific situation at hand in order to reach the best decision possible for a particular patient. This type of reasoning—from the general to the specific—is called **deductive reasoning.** Much of basic medical training centers around deductive reasoning.

If it has not happened already, at some point in your training you must ask yourself: How do we obtain the information about what happens in general? We are told, for example, that patients with hypertension eventually have strokes if their blood pressure is not controlled, but how did we obtain this information in the first place? Does the rule always hold? Are there exceptions? How long can the patient go untreated without having a stroke? Just how high can the blood pressure level be before the patient is in imminent danger?

These questions are answered by "experience." But how do we pyramid the knowledge we glean from experience so that we do not make the same mistakes over and over again? We save the information gathered from experience and refer to it to make better judgments as we are faced by the need to make new decisions. Moreover, we conduct experiments and comparative studies to focus on questions that arise in our work. We study a few patients (or experimental animals), and from what we observe we try to make rational inferences about what happens in general. This type of reasoning—from the specific subject(s) at hand to the general—is called **inductive reasoning**. This approach to medical research—pushing back the bounds of knowledge concerning human health—follows what is known as the **scientific method**, which has four basic steps:

1. Making observations—i.e., gathering data.
2. Generating a hypothesis—the underlying law and order suggested by the data.
3. Deciding how to test the hypothesis—what critical data are required?
4. Experimenting (or observing)—this leads to an inference that either rejects or affirms the hypothesis. If the hypothesis is rejected, then we go back to step 2. If it is affirmed, this does not necessarily mean it is true, only that in light of current knowledge and methods it appears to be so. The hypothesis is constantly refined and tested as more knowledge becomes available.

It would be easy to reach conclusions on the basis of observations, were it not for the variability inherent in virtually all biological data. One of the most common decisions a health professional must make is whether an observation on a patient should be considered normal or abnormal. Is a particular observation more typical of a person with disease or of a person without disease? Is the observation outside the range typically found in a healthy person? If the patient were examined tomorrow, would one obtain essentially the same observation? Obviously, observations such as blood pressure evaluations vary greatly, both at different times on the same patient and from patient to patient. Clinical decisions must be made with this variability in mind.

Inductive inference is a much more risky procedure than deductive inference. In mathematics, we start with a set of axioms. Assuming that these axioms are true, we use **deductive** reasoning to prove things with certainty. In the scientific method, we use **inductive** inference and can never prove anything with absolute certainty. In trying to generalize results based on a group of 20 patients, you might ask such questions as: If 20 additional patients were studied, would the results be very close to those obtained on studying the first 20 patients? If a different laboratory analyzed the blood samples, would the results be similar? If the blood samples had been stored at a different temperature, would the results be the same?

**A statistic** (plural: statistics) is an estimate of an unknown numerical quantity, such as the mean height of men age 20. **Statistics** (singular) is a science that deals with the collection, organization, analysis, interpretation, and presentation of information that can be stated numerically. **Biostatistics** is statistics applied to the biological sciences. Perhaps the most difficult aspect of statistics is the logic associated with inductive inferences, yet all scientific evidence is based on this type of statistical inference. The same logic is used, though not always explicitly, when a physician practices medicine: what is observed for a particular patient is integrated with what has previously been observed for a large group of patients to make a specific decision about that particular patient.

## REASONS FOR STUDYING BIOSTATISTICS

Many students ask: Why should I study biostatistics? or: Why would a health professional need to know statistics? The statement "If I need a statistician, I will hire one," is also common. But health professionals are frequently faced with data on which they must base clinical judgments. The reliability of support data plays a fundamental role in making good clinical decisions. You must be able to distinguish between discrepant data and routine variability. As a layman and as a practitioner, you will be bombarded daily with statistics. To make correct decisions based on the data you have, you must know where those data came from and how they were obtained; you must also know whether conclusions based on those data are statistically valid. Statistics are often misinterpreted. Disraeli is reputed to have said: "There are lies, damned lies, and statistics."

You must be able to understand and evaluate the medical literature in an intelligent manner. Unfortunately, many of the articles in the medical literature draw invalid conclusions because incorrect statistical arguments are used. Schor and Karten (1966) found that most analytical studies published in well-respected medical journals in 1964 were unacceptable in that the conclusions drawn were not valid in terms of the design of the experiment, the type of analysis performed, or the applicability of the statistical tests used. Unfortunately, things were only slightly better 15 years later: Glantz (1980) reported that about half of the articles published in medical journals that use statistics use them incorrectly. After you complete your training, you will be relying on the literature for new information that will change the way you practice your profession. It is important that you be able to read these articles critically. You will need to understand terms such as "p-value," "significance level," "confidence interval," "standard deviation," and "correlation coefficient," to mention just a few of the statistical terms that are now common in the medical literature. This book explains these concepts and gives you an explanation of statistics that will help you distinguish fact from fancy in everyday life—in newspapers and on television, and in making daily comparisons and evaluations.

The easy availability of computers means that increasingly sophisticated data analysis will be performed to advance medical science. The computer augments cerebral functions in the same way that other machines augment muscular functions: the computer can receive, store, select, evaluate, and transmit information. To understand the meaning of computer output, you will need to understand the principles underlying the computations performed by the computer. Remember, however, that no amount of computation can produce information that is not already inherently present in a set of data. Computer scientists refer to this principle as GIGO—Garbage In, Garbage Out!

Finally, you should have an appreciation of statistics so that you know when, and for what purpose, a statistician should be consulted in medical research.

# HOW CAN A STATISTICIAN HELP AN INVESTIGATOR?

Statistics is a vital component of the research process, from the earliest planning stages of a study to the final presentation of its results. The involvement of a statistician in all stages of a research project enhances the efficiency of the study and the scientific credibility of its results. If a study has been improperly planned or executed, no amount of statistical expertise can salvage its results. At the beginning of a study, the statistician's activities might include: (1) recommending study designs to meet the objectives and to increase the amount of information that can be obtained; (2) helping develop efficient data-collection forms that are easily processed; and (3) recommending ways to monitor the quality of the data as they are being collected. After the data have been collected and prepared for analysis, the statistician can: (1) recommend the most appropriate methods of analysis, and possibly do the analyses personally if the methods are sophisticated; (2) interpret the findings in understandable terms; and finally (3) review and contribute to the statistical content of any presentations and publications that report the results of the study. Statisticians may also be consulted to help evaluate published papers and manuscripts, or to help prepare sections on experimental design and analysis for grant applications.

# WHAT DOES THE STATISTICIAN EXPECT FROM AN INVESTIGATOR?

Because statistical consultation is a professional collaboration, statisticians should be included in research projects from their beginning. Also, because statistical design and analysis are time-consuming activities, statisticians should be informed well in advance of any deadlines. At the beginning of the study, you should: (1) show the statistician exactly where and how the data will be collected, preferably at the collaboration site; (2) describe the objectives of your study in detail, because they are essential in planning a study design that will extract all pertinent information as efficiently as possible; and (3) inform the statistician of any relevant limitations, such as availability of financing or personnel, which are all factors that must be considered in the study design. To clarify the terminology and basic concepts in your field of interest, you should provide the statistician with background information in the form of basic articles and book chapters in your area of research.

Once a strategy for your study has been jointly established, it must be followed carefully, and any necessary changes in procedures must be discussed before they are implemented. Even minor changes may affect the way data are analyzed, and in some cases could invalidate an entire study. Although certain statistical methods may partially correct mistakes in study design, this is not always possible; it is obviously expedient to avoid making such mistakes at all.

In return for substantial contributions to a study, the statistician could reasonably expect to be a co-author on publications reporting the results of the study, and should, of course, have the opportunity to review each publication before it is submitted. In the case of a minor contribution, an acknowledgment may be sufficient, but remember that it is as discourteous to acknowledge someone's help without that person's permission as it is to fail to make an appropriate acknowledgment. The question of how the statistician's contribution will be recognized should be brought up early in a project.

## SOME SPECIFIC EXAMPLES

We are all aware of the accuracy of projections made by the media in predicting the outcome of national elections before many people have even gone to the polls to cast their vote. This process is based on a relatively small but representative sample of the population of precincts throughout the country. It is a classic example of statistical inference—drawing conclusions about the whole population based on a representative sample of that population.

Several years ago the King Tut art exhibit was on display in New Orleans. During the last few days of the exhibition, people waited in line for hours just to enter to see it. On the very last day, the lines were exceptionally long and seemed to be moving terribly slowly. One ingenious man decided to estimate his expected waiting time as follows. He stepped off 360 paces (approximately 360 yards) from his position to the front of the line. He then observed that the line moved

10 paces in 15 minutes. He projected this to estimate a movement of 40 paces per hour or 360 paces in 9 hours. The man then decided that 9 hours was too long a period to wait in line. A man (one of the authors of this book) directly behind this fellow, however, decided to wait and stood in line 9½ hours before seeing the exhibit!

In 1948, a report appeared of what turned out to be a classic study of streptomycin and tuberculosis. Patients were assigned to a streptomycin group or a control group and assessed after 6 months. Fourteen of 52 patients in the control group died during the 6-month period, compared with 4 of 55 patients in the streptomycin group. Moreover, among those who lived, significantly more patients in the streptomycin group showed considerable improvement on radiological evaluation. It was inferred from this, and subsequent experience has provided even stronger evidence, that streptomycin is an effective treatment for tuberculosis.

Today most health experts believe that smoking is bad for the health—that it increases the risk of cancer and heart attacks, and has other deleterious effects on the body. Governments have taken actions to modify or suppress advertising by the tobacco industry, and to educate the public about the harmful effects of smoking. These actions were taken, however, only after many independent studies collected statistical data and drew similar conclusions.

## SUMMARY

1. Biostatistics deals with the collection, organization, presentation, analysis, and interpretation of biological information that can be stated numerically.

2. All data collected from biological systems have variability. The statistician is concerned with summarizing trends in data and drawing conclusions in spite of the uncertainty created by variability in the data.

3. Deductive reasoning is reasoning from the general to the specific. Inductive reasoning is drawing general conclusions based on specific observations.

4. The scientific method provides an objective way of formulating new ideas, checking these ideas with real data, and pyramiding find-

ings to push back the bounds of knowledge. The steps are: make observations, formulate a hypothesis and a plan to test it, experiment, and then either retain or reject the hypothesis.

5. An understanding of statistics will enhance your ability to interpret data, whether for the purpose of treating a particular patient or for drawing general conclusions from a research study, as well as enable you to distinguish fact from fancy in everyday life.

## FURTHER READING

Glantz, S. A. (1980) Biostatistics: How to Detect, Correct and Prevent Errors in the Medical Literature. Circulation 61:1–7. (You will gain more from this article if you read it after you have learned biostatistics.)

Huff, D. (1954) How to Lie with Statistics. New York: Norton. (Everyone who has not read this book should do so; you can read it in bed!)

Medical Research Council (1948) Streptomycin Treatment of Pulmonary Tuberculosis. British Medical Journal 2:769–782.

Schor, S. S. and Karten, I. (1966) Statistical Evaluation of Medical Journal Manuscripts. Journal of the American Medical Association 195:1123–1128.

## CHAPTER 2

# Populations and Samples

**Key Concepts:** cause and effect; confounding; target population, study population, study unit, census, parameter; probability sample: simple random sample, stratified random sample, systematic random sample; observational study: cohort/prospective study, case-control/retrospective study; experimental study: completely randomized, randomized blocks, split-plot, change-over/cross-over design, sequential design; factorial arrangement of treatments; clinical trial, placebo effect, blinding, double blinding.

## THE STUDY OF CAUSE AND EFFECT

Very early in your study of science you probably learned that if you put one green plant in an area exposed to sunlight and another green plant in a dark area, such as a closet, the plant in the dark area would turn yellow after a few days, whereas the one exposed to sunlight would remain green. This observation involves a simple experiment in which there are just two plants. Can we infer from it that we have a cause (light) and an effect (green color)? Will the same thing happen again? We might put several plants in sunlight, and several in a closet, and we might repeat the experiment on many different occasions, each time obtaining the same result. This would convince us that what we observed was not pure coincidence, but can we be sure that it is the light that causes the green color (or, conversely, darkness that causes the yellow color)? If the closet is always cooler than the sunlit areas, the color change could be simply due to the cooler temperature. Clearly such an experiment—in which temperature is not carefully controlled—cannot distinguish between lack of sunlight and cooler temperature as the cause of the color change. In this situation, we would say that the effect of light is **confounded** with that of temperature. Two factors in a study are said to be confounded when it is not possible to determine from the study which factor is causing the effect being investigated. In this chapter, we are going to discuss ways of designing studies with appropriate safeguards against confounding and other pitfalls, so that we can be more certain about making correct inferences about causes and effects.

Much of medicine is concerned with the study of disease—identifying the cause of disease, preventing disease by intervention, and treating disease to minimize its impact. Unfortunately, most diseases have a complex pathogenesis, and so it is not easy to describe the underlying process. In the simplest situation, factor A causes disease X. In more complex situations, disease X is caused by multiple factors, say factors A, B, C, and D, or factor A may cause more than one disease. Factor A may cause disease X only in the presence of factor B, so that if either factor A or factor B is present alone, no causative effect can be observed. In another instance, factor A could initiate disease X, but factor B could accelerate or promote the disease process once it is initiated. On the other hand, disease X could influence factor A, whereas factor A might have no influence on the disease. The time of exposure to possible causative factors is another consideration. Moreover, it is not always obvious that a disease is present, especially in the preclinical or early clinical stages. Some patients might have more than one disease at the same time, and these diseases might be associated with some of the causative factors. Because of these many possibilities it is essential that studies be carefully designed.

In any good scientific study the objectives will be clearly stated, including specific hypotheses to be tested and effects to be estimated. For example, the objectives might be to:

1. Identify a group of men with elevated serum cholesterol levels.
2. Reduce the serum cholesterol levels in these men by administering a treatment, say drug A.
3. Determine whether a reduction in serum cholesterol levels over a 5-year period reduces the risk of developing coronary heart disease.

4. Determine whether drug A has harmful side effects that out-weigh any reduction in the risk of coronary heart disease.

Thus the specific hypothesis to be tested is that using drug A over a 5-year period, to lower the serum cholesterol levels of men with elevated levels, reduces their risk of developing coronary heart disease. The specific effects to be estimated are the amount of reduced risk and the amount of harmful side effects.

At the outset, definitions of terms should be specified. For example, coronary heart disease might be defined as myocardial infarction or a history of angina, but this definition in turn requires definitions of myocardial infarction and angina. If the definitions of these terms are not clearly documented, it is easy for inconsistent use of the terms to enter into the study. If patients who are free of coronary heart disease are entered into the study, it might be more difficult to estimate the effects we are interested in. We also need to define what is meant by an "elevated" serum cholesterol level. This will determine which men will enter the study, and hence the men for whom any results of the study will have any relevance. We now turn to a general discussion of how the inferences we wish to make are limited by the manner in which our study units are selected (in the example we have just considered, each "study unit" is a man with an elevated serum cholesterol level).

## POPULATIONS, TARGET POPULATIONS, AND STUDY UNITS

One of the primary aims of statistics is to help draw objective conclusions that pertain to a larger group than the one for which data are available. You might want, for example, to compare two treatments for angina. Obviously, you cannot give the treatments to all patients with angina, but you can study a small group of such patients. You would then try to generalize your conclusions based on the results from that small group to a larger group of patients—perhaps even to most patients with angina. Similarly, it would not be practical to study all men with elevated serum cholesterol levels, but we would hope to learn something relevant to all such men on the basis of an experiment performed on a small group, or sample, of such men.

In these examples, the set of all angina patients, or the set of all men with elevated cholesterol levels (the word "elevated" being precisely defined), would be the **population** about which we wish to make inferences. The population is made up of **study units**, i.e., the units we study from the population—which in these examples are angina patients or men with elevated cholesterol levels, respectively. Although it seems obvious in these examples that we are studying people, and that these people make up specific populations, the situation is not always this clear. The population of interest may, for example, be made up of blood samples or tissue specimens, each such sample or specimen being a study unit. We often use animal models in our initial investigations of an area of human research, and so the study unit might be a rat, a hamster, or a dog.

The important point to remember is that in statistics, a population is the group of all study units about which a particular investigation may provide information. The study units make up the population, and the population about which we wish to make inferences determines what is meant by a study unit. Suppose, for example, that we are interested in the functioning of the muscle cells of patients with myotonic dystrophy. We might take two muscle biopsies from each of five such patients and divide each biopsy into three aliquots, or parts, making a total of 30 aliquots in all, on each of which we make a measurement. But, if the population about which we wish to make inferences is the population of patients with myotonic dystrophy, our sample contains only five study units—on each of which we have the measurements made on six aliquots. If we wish to make inferences about all the (theoretically) possible biopsies that could be taken from one particular patient, then each biopsy is a study unit and we have a sample of only two such units on any one patient. It is a common error to believe that increasing the number of measures taken on any one study unit is equivalent to increasing the number of study units sampled from a population. If our experiment with the green plants had involved only two plants—one in the light and one in the dark— then, since we were interested in making inferences about plants, only two study units were in the experiment. This number is not increased to 20 by noting that each plant has ten leaves and then recording the color of each leaf. We must carefully distinguish between multiple study units and multiple measures on a single study unit—a distinction that is intimately tied to the population about which we wish to make inferences.

We must also carefully distinguish between the **target population** and the **study population.** The target population is the whole group of study units to which we are interested in applying our conclusions. The study population, on the other hand, is the group of study units to which we can legitimately apply our conclusions. Unfortunately, the target population is not always readily accessible, and we can study only that part of it that is available. If, for example, we are conducting a telephone interview to study all adults (our target population) in a particular city, we do not have access to those persons who do not have a telephone. We may wish to study in a particular community the effect of drug A on all men with cholesterol levels above a specified value, but short of sampling all men in the community, only those men who for some reason visit a doctor's office, clinic, or hospital are available for a blood sample to be taken. Thus, we have a study population of accessible study units and a target population that includes both the study population and the inaccessible study units. Those study units that are not readily accessible may or may not have the same characteristics as those of the study population. Their exclusion from the study population means that inferences made about the study population need not necessarily apply to the target population.

There are many ways to collect information about the study population. One way is to conduct a complete **census** of the population by collecting data for every study unit in it. The amount of money, time, and effort required to conduct a complete census is usually unreasonable. A more practical approach is to study some fraction, or **sample,** of the population. If the sample is representative of the population, then inferences we make from the sample data about the population will be correct. The term **statistic** is used to designate an estimate computed from sample data, and the term **parameter** is used to designate a quantity that is characteristic of the population. If the sample is representative of the population, descriptive statistics will give accurate impressions of the corresponding parameters of the population. Since our interest is in estimating parameters and testing hypotheses about parameters of the population, special efforts should be taken to obtain a representative sample. Haphazard samples, or samples selected on the basis of being easy to collect, are rarely representative of the population. We will now describe methods of sampling a population that usually lead to representative samples, and that, in any case, allow us to make valid inferences.

## PROBABILITY SAMPLES AND RANDOMIZATION

In order to make the kinds of inferences we discuss in later chapters, we select well-defined **probability samples,** in which every unit in the population has a known probability of being included in the sample. The most elementary type of probability sample is the **simple random sample,** in which every study unit in the population is equally likely to be included in the sample. However, it is often better to take another kind of probability sample. Suppose, for example, we take a simple random sample of all individuals in a community, and it just happens, by chance, that there are no women in our sample. Sex might be an important factor in what we are studying, and if this is the case, we would have obtained a sample that may be seriously deficient. To overcome this possibility, we take a **stratified random sample**—a sample obtained by separating the population study units into nonoverlapping groups, called **strata,** and then selecting a simple random sample from each stratum. Thus, the population is first divided by gender (the two strata), and a random sample then taken from each stratum. In this way we can insure that each gender is represented in the sample in proportion to its distribution in the population. Similarly, we might stratify the population on the basis of age, socioeconomic status, health status, and so forth, before taking a simple random sample of a predetermined size from each stratum.

One approach to selecting a random sample is to put the name, number, or some other identification of each study unit in the population into a container, such as a hat, mix the identifications thoroughly, and then select units (i.e., identifications) one at a time until the required sample size has been attained. This procedure, however, has practical limitations. It may not be possible to obtain a thorough mixing, and it may be impractical to write down the identification of every unit in the population on a separate piece of paper. However, provided we have a list of these identifications (such as a list of all hospital patients who, during a period of time have a particular disease, or a list of all residents in a community), a sequence of random numbers can be used to pick the sample. Thus, if our list contains 10,000 names, a 1% random sample can be selected by obtaining a sequence of 100 random numbers between 1 and 10,000, and using those numbers to indicate the positions in the list of those persons who are to form the sample. Tables of random numbers have been published for this purpose. We would start at an arbitrary place in the table and then take as our random numbers the next 100 sets of four digits that appear successively in the table. The four digits 0000 would be interpreted as 10,000, and in the unlikely event that the same number is obtained twice, another number would be taken. There are also computer programs that generate pseudo-random numbers—they appear to be random, but are in fact produced by a well-defined numerical algorithm. The fact that they do not produce truly random numbers is usually of little consequence when a single sample is being selected for a particular study.

A simpler way to select a sample from a list is to take what is called **a systematic random sample.** Here we randomly select a number between 1 and k—where 1/k is the fraction of study units in the population we wish to have in our sample—and then select every k-th unit in the list. This type of design is often used in selecting patients from a large clinic on the basis of hospital charts. We might study every 1 in 20 charts, for example, simply selecting a random number between 1 and 20, say 16, and studying the 16th, 36th, 56th, and so forth, chart. The disadvantage of this kind of sample is that the order of the study units on the list may have a periodic pattern, in which case we may obtain an unrepresentative sample. Provided we can be sure that this is not the case, however, a systematic random sample is a conveniently simple way to obtain an approximately random sample.

Often the study units appear in groups, and so we sample clusters. The physicians working in a hospital could be considered a cluster. Thus, we could sample hospitals (clusters) and interview every physician working in each hospital selected in the sample. A **two-stage cluster sample** is obtained by first selecting a sample of clusters (stage one) and then selecting a sample of study units within each cluster (stage two). In this situation, we really have two different study units, since the clusters themselves can be considered as individual study units in the population of all such clusters. As we shall see in Chapter 10, there are special methods of analysis for this kind of situation.

## OBSERVATIONAL STUDIES

So far, we have considered the problem of sampling study units from the population, and we have seen that different ways of doing this, or different **sampling designs**, are possible. The sampling design is usually critical to the interpretation of **observational studies**, that is, studies in which the researcher merely "observes" the study units, making one or more measurements on each.

There are many types of observational studies, e.g., to determine the prevalence of disease, or to determine the population distribution of a given trait, such as blood pressure. Two types of observational studies that we shall discuss in more detail provide different epidemiological approaches to investigate the cause of disease. The first approach is to identify a sample of persons in whom the suspected cause is present and a second sample in whom the cause is absent, and then compare the frequency of development of disease in the two samples. Studies that proceed in this manner are called **cohort**, or **prospective, studies.** The two samples are called cohorts, and we observe them "prospectively" for the occurrence of disease. We might, for example, identify one cohort of smokers and one cohort of nonsmokers, and then observe them for a period to determine the rate of development of lung cancer in the two groups.

The other approach is to identify a sample of patients who have the disease of interest (**cases**) and a second sample of persons who do not have the disease (**controls**), and then compare the frequency of possible causes of the disease in the two groups. These are called **case-control**, or **retrospective studies.** They are retrospective in that we "look back" for possible causes. Thus we might identify a group of lung-cancer patients and a group of lung-cancer–free subjects, and then study their past history of smoking.

Thus, the difference between cohort and case-control studies is in the selection of the persons (the study units) for study. In a cohort study, persons are selected who are initially free of disease, and we determine disease frequency over some period of time, in the presence or absence of factors suspected of being associated with causing the disease. On the other hand, in a case-control study, persons are selected on the basis of presence or absence of the disease and we determine the frequency of possible causative factors in their past histories. In both types of observational studies there is always the possibility that confounding factors are present. It is known, for example, that smokers tend to drink more coffee than nonsmokers. Thus, any association found between smoking and lung cancer could in theory merely reflect a causal link between coffee drinking and lung cancer. Provided the potential confounding factors can be identified, methods of analysis exist to investigate this possibility. The major drawback to observational studies is that there may be unidentified confounding factors.

In cohort studies, persons are usually selected at the time the study is started and then followed over time. Existing records alone, however, can be used to conduct a cohort study. The groups are established on the basis of possible causative factors documented at an early point in the records, and disease frequency is then established over a subsequent period in the same existing records (e.g., work records at an industrial plant). Such a study, even though it is conducted retrospectively, is called a **historical cohort**, or **historical prospective**, study. The important point is that the outcome (disease or not) is determined prospectively in relation to the time in the record at which the samples are chosen.

The source of cases in most case-control studies is provided by the patients with the disease of interest who are admitted to a single hospital or to a group of hospitals during a specific interval of time. The controls are often patients admitted to the same hospitals during the same interval of time for reasons other than and unrelated to the disease under study. Alternatively, instead of such hospitalized controls, an effort may be made to obtain population-based controls. The type of controls used can play a critical role in the interpretation of the results, since hospitalization **per se** (regardless of a person's status with respect to the disease being studied) may be a confounding factor related to the putative causative factors.

To obtain a group of controls that is comparable to the cases, the controls are often matched to the cases with respect to extraneous variables. Thus, for example, as a particular case is entered into the study, one or more persons of the same sex and race, and of similar age and socioeconomic status, are identified and entered into the control group. We say that the cases and controls have been **matched** for sex, race, age (within a specified interval, e.g., within five years) and socioeconomic status. A **matched pairs** design, implying one control matched to each case, is common.

The case-control study design is frequently used to explore simultaneously a number of possible causes of disease. Such a study can usually be conducted from existing records, and is therefore relatively inexpensive. Furthermore, it can be conducted in a reasonably short time. Often the purpose of a case-control study is to learn enough about the causes of a disease of interest to narrow the range of possibilities. Then a prospective study, with a small number of clearly defined hypotheses, can be planned and conducted.

## EXPERIMENTAL STUDIES

In **experimental studies**, the researcher intervenes ("experiments") in some way to affect the manner in which the study units (in this case often called "experimental units") respond. The study units are given some stimulus, or treatment, and then a response is observed. For example, our study units could be patients with the common cold, and the treatment could be a drug that we believe may be a cure for the common cold. We give the drug to a sample of such patients, and two weeks later observe that virtually none of them has a cold any more. This is, in fact, an experimental study, but does it allow us to conclude that the drug is a cure for the common cold? It clearly does not, because we would expect the same result even if the patients had not received the drug. We all know that common colds tend to go away after a couple of weeks whether or not any drugs are taken. This example illustrates the very simple and obvious principle that if a result follows a particular course of action, it does not mean that the course of action necessarily causes the result. Studies of treatment efficacy are often plagued by ignoring this simple principle—that in many cases, time alone is sufficient to "cause" a cure or an amelioration of a disease.

For time to be eliminated as a possible confounding factor, experimental studies must have at least two treatment groups. The groups are compared with each other after the same amount of time has elapsed. If there are two groups, one will be given the treatment being investigated, and the other, the control group, may be given no treatment. We say there are two different treatment groups, since "no treatment" is in itself a type of treatment. Thus, we say there are two possible treatments: drug therapy versus no drug therapy. Or, if it would be unethical not to treat the patient at all, the two treatments could be a new drug we are investigating and a control treatment that has been (until now) the standard treatment for such patients. There may be more than two groups—one being given the new treatment and two or more others being given two or more other standard treatments for the disease. The way in which the study units are assigned to the various treatment groups, together with any relationships there may be among the different treatments, determines what is known as the **experimental design.**

Careful randomization is an essential component of any sound experimental design. Ideally, all of the study units in an experiment should be obtained from the target population using an element of randomization. But this is usually impractical, and the sample of study units used is often not at all representative of the target population. Nevertheless, randomization plays an important role in the allocation of study units to the various treatments of experimental studies. Suppose, for example, that patients with a specific disease are assigned to one of four treatments, and that after a suitable period of observation, the treatment groups are compared with respect to the response observed. Although the patients studied might be representative of those seen at a particular clinic, they are unlikely to be representative of all patients with the disease of interest (the target population). But if the patients are assigned to the treatments in such a way that each patient is equally likely to receive any one of them, a fair comparison of the four treatments is possible. Thus, the patients should be randomly assigned to the treatment groups, as this will enhance the chances of a fair distribution of the good- and poor-risk patients among the four groups. If 32 patients are available, they can each be given a number, the 32 numbers written on separate pieces of paper, shuffled, and then randomly sorted into four treatment piles.

A table of random numbers could also be used for the same purpose. It cannot be stressed too much that without a proper randomization procedure, biases—wittingly or unwittingly—are almost bound to result. If we had 32 mice in a cage to be assigned to four different treatments, it would not be sufficient simply to reach in and pick out eight "at random" for the first treatment, eight "at random" for the second, and so on. An obvious bias occurs with this procedure in that the last group will consist of those mice that are less easy to catch, and the first group those easiest to catch. Random allocation of study units to the treatment groups is an automatic safeguard against possible confounding factors. This is an essential difference between any observational study and a well-conducted experimental study (i.e., an experimental study in which there is random allocation of study units to treatments).

Many types of experimental designs are available, all involving an appropriate element of randomization. The main reason for choosing one design over another is to save the experimenter money, time, and effort, while at the same time insuring that the objectives of the study can be met. It will be helpful to review very briefly a few of the most common experimental designs.

A **completely randomized design** is one in which each study unit has the same probability of being assigned to any treatment group under consideration. Thus, if three treatments, A, B, and C, are being investigated, then each study unit would have a $1/3$ probability of being assigned to any one of the three treatment groups.

A **randomized blocks design** is one in which the study units are first divided into nonoverlapping "blocks" (sets of units or strata) and then randomly assigned to the treatment groups separately within each block. Consider, for example, an experiment in which a group of patients is separated by gender to form two blocks. The male patients are randomly assigned to the treatment groups (using a completely randomized design approach), and then the female patients are separately randomly assigned to the treatment groups. In this way, if patients of one gender always tend to give a better response, this "sex effect" can be "blocked out" in the sense that treatments are compared within the blocks of male and female patients. Provided the treatment **differences** are similar among the male and among the female patients, it is possible to pool the results from the two sexes to obtain an overall comparison of treatments that is not obscured by any sex effect.

The **split-plot design** involves randomly assigning units ("whole plots") to treatment groups and then randomly assigning subunits ("split plots") to a second kind of treatment group. (The use of the term "plot" arose from agricultural experiments in which each experimental unit was a plot in a field.) For example, patients (whole plots) with advanced diabetes might be randomly assigned to different treatments for diabetes and then their eyes (the split plots) might be randomly assigned to different treatments to correct eyesight.

The **change-over,** or **cross-over, design** is used, especially in animal and human studies, in an effort to use the study unit (patient) as his or her own control. In this design, units are randomly assigned to one treatment for a period, and then switched to another treatment for a second period. For example, if two treatments, say A and B, are being investigated, one group of patients would be assigned to treatment A for, say two weeks, and then to treatment B for two weeks. A second group of patients would be assigned to treatment B for two weeks followed by treatment A for two weeks. In this design, a residual or carry-over effect of the treatment given in the first period may affect the results found in the second period. In an attempt to prevent this, a rest period is sometimes given between the two treatments. Provided there is no carry-over effect, or if the carry-over effect is the same when B is followed by A as when A is followed by B, there is no problem in using all the data to estimate the true difference between the effects of the two treatments. If this is not the case, difficulties arise in analyzing the results of this kind of experimental design.

In many clinical studies, patients are recruited into the study over weeks, months, or even years. Whenever there is a lapse in time between the study of successive experimental units, a **sequential design** may be used. In such a design, the data are analyzed periodically as they are collected, and the result of each analysis determines whether to continue recruiting patients into the study or whether the study should be terminated and a decision made, on the basis of the data gathered so far, as to which treatment is best. It is also possible to use the result of each analysis to determine the probability of assigning the next patient to each of the treatments. In this way, as evidence accumulates that a particular treatment is best, the next patient recruited into the study has greater probability of receiving that treatment. This strategy is called "playing the winner."

So far, we have discussed that aspect of experimental design that concerns the way in which the experimental units are assigned to the different treatments. We have stressed that randomization must be involved at this step if we are to avoid biases and be able to make valid inferences. Another aspect of experimental design concerns the choice of the different treatments to investigate. We may be interested in the effects of three different doses of drug A—call these treatments $A_1$, $A_2$, and $A_3$. We may also be interested in the effects of two different doses of drug B, say $B_1$ and $B_2$, as treatments for the same disease. Furthermore, we may be interested in investigating whether there is any advantage in using these drugs in combination. We could set up separate experiments to investigate each of these three questions. But it is more economical to investigate the three questions simultaneously in a single experiment that has a **factorial arrangement** of the treatments. By this we mean that there are two or more factors of interest (two in our example—drug A and drug B), each at two or more levels (three levels of drug A and two levels of drug B in our example), and that our treatments comprise all possible combinations of different levels of each factor. Thus, in our example, there would be $3 \times 2 = 6$ different treatments, which we can label $A_1B_1$, $A_1B_2$, $A_2B_1$, $A_2B_2$, $A_3B_1$, and $A_3B_2$.

There are many other general aspects of experimental design, only a few of which can be briefly mentioned here. Laboratory techniques should be refined to minimize, or perhaps even eliminate entirely, sources of extraneous variability, such as observer biases, measurement errors, and instrument variability. Where possible, large sources of variability should be used as a basis for defining "blocks" in a randomized blocks experiment. Sometimes it may be possible to measure, but not control with any accuracy, factors that could be important sources of variability. In a biochemical experiment in which each study unit is a reaction mixture, for example, it may be possible to control the temperature of each mixture to within 2°C, but no more accurately than that. On the other hand, it might be possible to measure the actual temperature attained in each mixture with great accuracy. If such small temperature changes could be critical to the outcome, then the temperature of each mixture should be recorded. Uncontrolled variables that may have an important bearing on the outcome are called **concomitant variables,** and a statistical technique called the analysis of covariance makes allowances for them; this is briefly discussed in Chapter 10. The precision of experimental results can often be greatly improved if appropriate concomitant variables are measured at the time of a study.

Finally, an important consideration in any investigation is the number of study units to include. The more study units observed, the more reliable our conclusions will be; but we should like to obtain reliable results with a minimum amount of money, time, and effort. As we shall see in Chapter 7, statistical methods are available for estimating the number of study units required once a study design has been chosen. We will now turn to special considerations of experimental design when the study units are human subjects.

**Clinical trials** are experimental studies that involve people as study units. Clinical trials have come to play a major role in deciding the efficacy of new drugs and other treatments as they become available. Early investigations of new treatments tend to focus on animal studies, but the ultimate evaluation involves a clinical trial.

In the case of a clinical trial, randomization tends to provide a good distribution of both poor- and good-risk patients in all treatment groups. Obviously, if one treatment group were assigned only good-risk patients and the other only poor-risk patients, a subsequent comparison of treatment effects would be biased. But even if we achieve a perfect randomization to the different treatments, it is still possible to misinterpret the effective differences among the treatments. If one group of patients is given an injection of a drug and the other is not, we cannot tell whether the difference in outcome is caused by the drug itself or by the act of injection. (Recall the example of the green plant that turns yellow in the closet: without a proper control we cannot be sure whether this effect is caused by less light or less heat.) A better plan would be to inject the control patients in the same manner, with a similar fluid, but one that does not contain the active drug. In this way the act of injection is no longer a confounding variable.

To enhance objectivity in evaluating treatments in clinical trials, the patient and/or the evaluating clinician are often not told which treatment the patient is receiving. This is especially important if one of the treatments is the administration of an inactive treatment (such as an injection or a pill containing no active drug), called a **placebo.** It is not unusual to observe an improvement in a group of patients on placebo therapy (a "placebo effect") when the patients do not know they are receiving a placebo; however, if they know they are receiving a placebo the effect is destroyed. Withholding information about which treatment is being used is called **blinding;** when both the patient and the evaluating clinicians are blinded, it is called **double blinding.**

Researchers involved in clinical trials often go to great lengths to try to enhance adherence (compliance) to a treatment regimen. If the treatment is a one-time treatment, adherence is not a problem; but if it is a treatment in which the patient must daily take a prescribed drug, then adherence is often poor. Frequent contact (e.g., once a week) between the physician and the patient can enhance adherence, but can also be prohibitively expensive. It may be helpful to provide motivational programs including educational lectures and group discussions. Adherence can be monitored by counting the number of pills remaining in the container at each visit, or by using specially designed dispensers.

An important consideration in clinical trials is ethics. If a physician believes that one treatment is better than another, can that physician ethically assign patients to treatments in a random fashion? On the other hand, is it ethical not to conduct a clinical trial if we are not sure which treatment is best? If the risks of side effects are great, can the physician ethically prescribe a drug? If early in the trial it becomes evident that one treatment is preferable to another, can we continue to randomly assign patients to all the treatments? Questions such as these must be addressed in all clinical trials. Furthermore, the purpose of the study and the possible risks must be described in lay terms to the patient, who must then sign an "informed consent" form agreeing to participate in the study. But the patient must be given the option of leaving the study at any time, and if this option is exercised, care must be taken in analyzing the results of the trial to insure that it introduces no serious biases.

# SUMMARY

1. The investigation of cause and effect requires well-designed studies. That B follows A does not imply that A causes B, since confounding factors may be present.

2. Study units are sampled from a study population, which is usually only part of the target population of interest. The set of all possible study units makes up the study population. The study population about which we make inferences determines how the study units are defined. Multiple measurements made on a study unit do not increase the sample size.

3. Selection of samples from a population using an element of randomization allows one to draw valid inferences about the study population. If the population is heterogeneous, better representation is obtained by use of a stratified random sample. A systematic random sample is a convenient approximation to a random sample.

4. Two types of observational studies are used to investigate the causes of disease. In cohort, or prospective, studies, samples are chosen on the basis of the presence or absence of a suspected cause, and then followed over time to compare the frequency of disease development in each sample. A cohort, or prospective, study is termed historical if it is conducted totally on the basis of past records. In case-control, or retrospective, studies, samples are chosen on the basis of presence or absence of disease, and compared for possible causes in their past. The choice of controls in case-control studies is critical: they may be hospital- or population-based; matching for demographic and other factors is usually desirable.

5. In experimental studies, there is intervention on the part of the researcher, who subjects the study units to treatments. Randomization by an approved method in the allocation of study units to the different treatment groups provides a safeguard against possible confounding factors, so that valid inferences are possible.

6. In a completely randomized design, each study unit has equal probability of being assigned to any treatment. Heterogeneity among study units can be "blocked out" in a randomized blocks design. A split-plot design allows study units, after being assigned to treatments at a primary level, to be divided into subunits for assignment to treat-

ments at a secondary level. In a change-over, or cross-over, design, each study unit is subjected to two (or more) treatments over time and comparisons among treatments are made within study units. This design can lead to difficulties in the analysis and/or interpretation if there are carry-over effects from one period to the next. In a sequential design, the data are analyzed periodically to determine whether and/or how to continue the study. A factorial arrangement of treatments is one in which the treatments comprise all possible combinations of different levels of two or more factors.

7. In all experiments, extraneous sources of variability should be kept to a minimum or blocked out. Any remaining variables that could have a large effect on the results but cannot be blocked out should be measured during the course of the experiment. Such concomitant variables can be used in the analysis to increase the precision of the results.

8. Clinical trials are experimental studies in which the study units are people. They are used to judge the efficacy of new drugs and other treatments. Great care is needed in the choice of the control or comparison (in view of the commonly found placebo effect) and in monitoring adherence to a regular regimen. Either the physician or the patient may be blinded as to which treatment is being used. Ideally, both are blinded, in a "double-blinded" trial. Ethical considerations play an important role in the design of clinical trials.

## FURTHER READING

Altman, D.G. (1980) Statistics and Ethics in Medical Research: Study Design. British Medical Journal 281:1267–1269. (This is a succinct overview of the relation between design and ethics in observational studies and clinical trials.)

Marks, R.G. (1982) Designing a Research Project: The Basics of Biomedical Research Methodology. Wadsworth. (This is a book on the practical aspects of design for a novice researcher.)

## PROBLEMS

1. A physician decides to take a random sample of patient charts for the last five years at a large metropolitan hospital to study the

frequency of cancer cases at that hospital. He estimates the number of charts to be 10,000 and decides to take a 5% sample (i.e., a sample of 500 charts). He decides to randomly select a number between 1 and 20 (e.g., suppose the number turned out to be 9), and then study every 20th chart beginning with that number (in this example, charts 9, 29, 49, 69, . . .). This is an example of a sample design known as a

  A. two-stage random sample
  B. stratified random sample
  C. systematic random sample
  D. random cluster sample
  E. simple random sample

2. A researcher decides to conduct a clinical trial using 40 patients. She carries out her treatment assignment by a well-defined method of randomly allocating 20 patients to group I and 20 patients to group II. After four weeks on treatment A, patients in group I are taken off the treatment for four weeks, then given treatment B for an additional four weeks. Similarly, patients in group II are given treatment B for four weeks, no treatment for four weeks, and then treatment A for four weeks. The design used in this study is called

  A. a stratified design
  B. a sequential design
  C. a completely randomized design
  D. a randomized block design
  E. a change-over design

3. Studies in which the study units are stimulated in some way and the researcher observes a response are called

  A. observational studies
  B. prospective studies
  C. sample surveys
  D. experimental studies
  E. retrospective studies

4. Pregnant women were recruited into a drug trial during the 28th week of pregnancy. They were allocated at random and double-blinded to placebo or active treatment groups. This could be best defined as

A. a sample survey
B. an experimental study
C. a retrospective study
D. a case-history study

5. In a study of the cause of lung cancer, patients who had the disease were matched with controls by age, sex, place of residence, and social class. The frequency of cigarette smoking in the two groups was then compared. What type of study was this?
A. Sample survey.
B. Experimental study.
C. Retrospective study.
D. Clinical trial.
E. Prospective study.

6. In a double-blind, randomized trial of the effectiveness of a drug in the treatment of ulcers, patients were randomly assigned to either an active or a placebo group. Each person was followed up for six weeks and evaluated as showing: (1) significant improvement; or (2) no significant improvement. The purpose of the double-blind aspect was to
A. obtain a representative sample of the target population
B. achieve a good distribution of good- and poor-risk patients in the two groups
C. guard against observer bias
D. eliminate the necessity for a statistical test
E. insure proper randomization

7. A physician decides to take a random sample of hospitals in a large metropolitan area. From each hospital included in the sample he takes a random sample of house-staff physicians. He interviews the physicians to determine where they attended medical school. The sample design used in this study is an example of a
A. systematic random sample
B. stratified random sample
C. simple random sample
D. two-stage random sample

8. A study was undertaken to compare results of the surgical treatment of duodenal ulcer. A total of 1,358 patients who met the study criteria were randomly assigned to one of four surgical procedures. The purpose of the randomization was to
   A. insure that the double-blind aspect of the study was maintained
   B. obtain the unbiased distribution of good- and poor-risk patients in all treatment groups
   C. achieve the same number of patients in each operation
   D. guarantee that the study group was a representative sample of the general population
   E. remove the poor-risk patients from the study

9. Investigators who use non-randomized controls in clinical trials often argue that their controls are satisfactory because the distribution of prognostic factors in the control and experimental groups is similar before therapy. It is better to randomize because
   A. a placebo effect is easier to detect in randomized trials
   B. randomization tends to magnify the differences between placebo and treatment groups
   C. many important prognostic factors that were not considered may lead to bias
   D. it is easier to maintain blinding in randomized trials
   E. adherence is better in randomized trials

10. Studies involving patients randomly assigned to treatment groups and then observed to study response to treatment are called
   A. retrospective studies
   B. case-control studies
   C. observational studies
   D. clinical trials
   E. sample surveys

## CHAPTER 3

# Descriptive Statistics

**Key Concepts:** interval, ordinal and nominal scale; continuous and categorical or discrete data; table, frequency distribution; histogram, bar-graph, frequency polygon, cumulative plot, scatter plot (scatter diagram), tree diagram; proportion, percentage, rate; prevalence, incidence; relative risk, odds ratio, attributable risk; sensitivity, specificity, predictive values; measures of central tendency: mean, median, mode; measures of spread: range, interquartile range, variance, standard deviation, coefficient of variation; skewness, kurtosis.

## WHY DO WE NEED DESCRIPTIVE STATISTICS?

We stated in Chapter 1 that a statistic is an estimate of an unknown numerical quantity. A descriptive statistic is an estimate that summarizes a particular aspect of a set of observations. Descriptive statistics allow one to obtain a quick overview, or "feel," for a set of data without having to consider each observation, or datum, individually. (Note that the word "datum" is the singular form of the word "data"; strictly speaking "data" is a plural noun, although, like "agenda," it is commonly used as a singular noun, especially in speech.)

In providing medical care for a specific patient, a physician must consider: (1) historical or background data; (2) diagnostic information; and (3) response to treatment. These data are kept in a patient chart that the physician reviews from time to time. In discussing the patient with colleagues, the physician summarizes the chart by describing the atypical data in it, which would ordinarily represent only a small fraction of the available data. To be able to distinguish the atypical data, the physician must know what is typical for the population at large. The descriptive statistics we discuss in this chapter are tools that are used to help describe a population. In addition, as we discussed in Chapter 2, researchers at universities and pharmaceutical companies conduct studies on samples of patients, and need to report their general findings to the medical community. Descriptive statistics are used to summarize the results of such studies in a succinct way. Tables and graphs are also useful in conveying a quick overview of a set of data, and in fact, tables and graphs are often used for displaying descriptive statistics. We therefore include a brief discussion of them in this chapter. First, however, we consider the different kinds of data that may need to be described.

## SCALES OF MEASUREMENT

We are all familiar with using a ruler to measure length. The ruler is divided into intervals, such as centimeters, and this is called an **interval** scale. An interval scale is a scale that allows one to measure all possible fractional values within an interval. If we measure a person's height in inches, for example, we are not restricted to measures that are whole numbers of inches. The scale allows such measures as 70.75 or 74.5 inches. Other examples of interval scales are the Celsius scale for measuring temperature and any of the other metric scales. In each of these cases the trait that is being measured is **quantitative,** and we refer to a set of such measurements as **continuous** data. Height, weight, blood pressure and serum cholesterol levels are all examples of quantitative traits that are commonly measured on interval scales. The number of children in a family is also a quantitative trait, since it is a numerical quantity; but it is not measured on an interval scale, nor does a set of such numbers comprise continuous data. Only whole numbers are permissible, and such data are called **discrete.**

Sometimes we classify what we are measuring only into broad categories. For example, we might classify a patient as "tall," "medium," or "short," or as "hypertensive," "normotensive," or "hypotensive." The trait is then **qualitative,** and such measurements also give rise to **discrete,** or **categorical,** data consisting of the counts, or numbers of individuals, in each category. There are two types of categorical data, depending on whether or not there is a natural sequence in which we can order the categories. In the examples just given, there is a natural order: "medium" is between "tall" and "short," and "normotensive" is between "hyptertensive" and "hypotensive." In this case the scale of measurement is called **ordinal.** The number of children in a family is also measured on an ordinal scale. If there is no such natural order, the scale is called **nominal,** with the categories having names only, and no sequence being implied. Hair color, for example (e.g., "brown," "blond," or "red"), would be observed on a nominal scale. Of course the distinction between a nominal and an ordinal scale may be decided subjectively in some situations. Some would argue that when we classify patients as "manic," "normal," or "depressed," this should be considered a nominal scale, while others say that it should be considered an ordinal one. The important thing to realize is that it is possible to consider categorical data from these two different viewpoints, with different implications for the kinds of conclusions we might draw from them.

## TABLES

Data and descriptive statistics are often classified and summarized in tables. The exact form of a table will depend on the purpose for which it is designed as well as on the complexity of the material. There are no hard and fast rules for constructing tables, but it is best to follow a few simple guidelines to be consistent and to insure that the table maintains its purpose:

1. The table should be relatively simple and easy to read.
2. The title should be clear, concise, and to the point; it should indicate what is being tabulated.
3. The units of measure for the data should be given.
4. Each row and column, as appropriate, should be labeled concisely and clearly.
5. Totals should be shown, if appropriate.
6. Codes, abbreviations, and symbols should be explained in a footnote.
7. If the data are not original, their source should be given in a footnote.

**TABLE 3.1.** Fasting Serum Triglyceride Levels (mg/100 ml) of 30 Male Medical Students

| | | | | |
|---|---|---|---|---|
| 45 | 46 | 49 | 54 | 55 |
| 61 | 67 | 72 | 78 | 80 |
| 83 | 85 | 86 | 88 | 90 |
| 93 | 99 | 101 | 106 | 122 |
| 123 | 124 | 129 | 151 | 165 |
| 173 | 180 | 218 | 225 | 287 |

Tables 3.1 and 3.2 are two very simple tables that display data we shall use, for illustrative purposes, later in this chapter. Table 3.1 is the simplest type of table possible. In it is given a set of "raw" data, the serum triglyceride values of 30 medical students. There is no special significance to the rows and columns, their only purpose being to line up the data in an orderly and compact fashion. Note that, in addition, the values have been arranged in order from the smallest to the largest. In this respect the table is more helpful than if the values

**TABLE 3.2.** Frequency Distribution of Fasting Serum Cholesterol Levels (mg/100 ml) of 1,000 Male Medical Students

| Cholesterol Level (mg/100 ml) | Number of Students |
|---|---|
| 90–100 | 2 |
| 100–110 | 8 |
| 110–120 | 14 |
| 120–130 | 21 |
| 130–140 | 22 |
| 140–150 | 28 |
| 150–160 | 95 |
| 160–170 | 102 |
| 170–180 | 121 |
| 180–190 | 166 |
| 190–200 | 119 |
| 200–210 | 96 |
| 210–220 | 93 |
| 220–230 | 35 |
| 230–240 | 30 |
| 240–250 | 23 |
| 250–260 | 15 |
| 260–270 | 7 |
| 270–280 | 3 |
| 280–290 | 1 |
| Total | 1,000 |

had simply been listed in the order in which they were determined in the laboratory. There is another kind of table in which the rows and columns have no special significance, and in which, furthermore, the entries are never ordered. This is a table of random numbers, which we referred to in Chapter 2.

The simplest type of descriptive statistic is a count, such as the number of persons with a particular attribute. Table 3.2 is a very simple example of how a set of counts can be displayed. Each of the observed 1,000 cholesterol levels occurs in just one of the interval classes, even though it appears that some levels (e.g., 100 and 110 mg/100 ml) appear in two consecutive classes. Should a value be exactly at the borderline between two classes, it is included in the lower class. This is sometimes clarified by defining the intervals more carefully (e.g., 90.1 – 100.0, 100.1 – 110.0, etc.). Age classes are often defined as 0–9 years, 10–19 years, etc. It is understood that the 10–19-year class, for example, contains all the children who have passed their 10th birthday but not their 20th birthday. Note that in Table 3.2 some of the information inherent in the original 1,000 cholesterol values has been lost, but for a simple quick overview of the data this kind of table is much more helpful than would be a table, similar to Table 3.1, that listed all 1,000 values.

## GRAPHS

The relationship between numbers of various magnitudes can usually be seen more quickly and easily from graphs than from tables. There are many types of graphs but the basic idea is to provide a sketch that quickly conveys general trends in the data to the reader. The following guidelines should be helpful in constructing graphs:

1. The simplest graph consistent with its purpose is the most effective. It should be both clear and accurate.
2. Every graph should be completely self-explanatory. It should be correctly and unambiguously labeled with title, data source if appropriate, scales, and explanatory keys or legends.
3. Whenever possible, the vertical scale should be selected so that the zero line appears on the graph.

4. The title is commonly placed below the graph.
5. The graph generally proceeds from left to right and from bottom to top. All labels and other writing should be placed accordingly.

One particular type of graph, a **histogram**, often provides a convenient way of depicting counts. Two examples of histograms, relating to the data in Tables 3.1 and 3.2, are shown in Figures 3.1 and 3.2. The points you should note about histograms are:

1. They are used for data measured on an interval scale.
2. The visual picture obtained depends on the width of the class interval used, which is to a large extent arbitrary. A width of 10 mg/100 ml was chosen for Figure 3.1, and a width of 20 mg/ 100 ml for Figure 3.2. It is usually best to choose a width that results in a total of 10 to 20 classes.
3. If the observations within each class interval are too few, a histogram gives a poor representation of the distribution of counts in the population. Figure 3.2 suggests a distribution with several peaks, whereas a single peak would most likely have been

**FIGURE 3.1** Histogram of 1,000 fasting serum cholesterol levels (from Table 3.2).

**FIGURE 3.2** Histogram of 30 fasting serum triglyceride levels (from Table 3.1).

found if 1000 triglyceride values had been used to obtain the figure. More observations per class interval could have been obtained by choosing a wider interval, but less than 10 intervals gives only a gross approximation to a distribution.

A **bar-graph** is very similar to a histogram, but is used for categorical data. It may illustrate, for example, the distribution of the number of cases of a disease in different countries. It would look very similar to Figures 3.1 and 3.2, but, since the horizontal scale is not continuous, it would be more appropriate to leave gaps between the vertical rectangles or "bars." Sometimes the bars are drawn horizontally, with the vertical scale of the graph denoting the different categories. In each case, as also in the case of a histogram, the length of the bar represents either a frequency or a relative frequency, sometimes expressed as a percentage.

A **frequency polygon** is also basically similar to a histogram, and is used for continuous data. It is obtained from a histogram by joining the mid-points of the top of each "bar." Drawn as frequency polygons, the two histograms in Figures 3.1 and 3.2 look like Figures 3.3 and 3.4. Notice that the polygon meets the horizontal axis whenever there is a zero frequency in an interval—in particular this occurs at the two ends of the distribution. Again the vertical scale may be actual frequency or relative frequency, the latter being obtained by dividing each frequency by the total number of observations; we have chosen to use relative frequency. A frequency polygon is an attempt to obtain a better approximation, from a sample of data, to the smooth curve that would be obtained from a large population. It has the further advantage over a histogram of permitting two or more frequency polygons to be superimposed in the same figure with a minimum of crossing lines.

A **cumulative plot** is an alternative way of depicting a set of quantitative data. The horizontal scale (abscissa) is the same as before, but the vertical scale (ordinate) now indicates the proportion of the observations less than or equal to a particular value. A cumulative plot of the data in Table 3.2 is presented in Figure 3.5. We see in Table 3.2, for example, that $2 + 8 + 14 + 21 + 22 + 28 + 95 = 190$ out of the 1,000 students have serum cholesterol levels less than or equal to 160 mg/100 ml, and so the height of the point above 160 in Figure 3.5 is

**FIGURE 3.3** Relative frequency polygon corresponding to Figure 3.1.

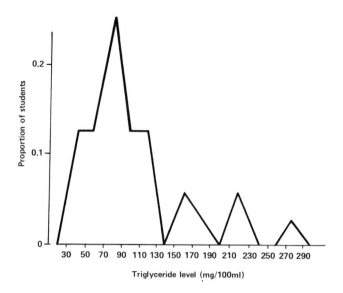

FIGURE 3.4 Relative frequency polygon corresponding to Figure 3.2.

[190]/[1,000], or 0.19. We could similarly draw a cumulative plot corresponding to the histogram of the 30 triglyceride values (Figure 3.2), but one of the great advantages of the cumulative plot is that it does not require one to group the data into interval classes, as does a histogram. In a cumulative plot every single observation can be depicted, as illustrated in Figure 3.6 for the data in Table 3.1. It is clear from that table that 1 out of 30 values is less than or equal to 45, 2 out of 30 less than or equal to 46, 3 out of 30 are less than or equal to 49, and so forth. So we can make $1/30$ the ordinate at 45, $2/30$ the ordinate at 46, $3/30$ the ordinate at 49, and so forth, up to $30/30 = 1$ as the ordinate at 287. However, the purpose of the cumulative plot is to approximate the continuous curve we would obtain with a much larger set of numbers. If more observations were included, one of them might possibly be larger than any of the values in Table 3.1. For this reason it is customary to make the ordinate at the largest data point (287 in this instance) somewhat less than unity. One convenient way of doing this is to use one more than the total number of observations as the divisor. Thus the ordinates for the data in Table 3.1 are depicted in

**FIGURE 3.5** Cumulative plot of the data in Table 3.2.

**FIGURE 3.6** Cumulative plot of the data in Table 3.1.

Figure 3.6 as $1/31$ at 45, $2/31$ at 46, $3/31$ at 49, up to $30/31$ at 287. Note that a cumulative plot results in a much smoother curve than the histogram (Figure 3.2), and that all the information in the original table is retained.

Many other types of graphs are possible, but only two more will be mentioned here. The first, the **scatter plot**, or **scatter diagram**, is an effective way of illustrating the relationship between two measures. In it, every point represents a pair of values, such as the values of two different measures taken on the same person. Thus, in the scatter plot depicted in Figure 3.7, every point represents a triglyceride level taken from Table 3.1, together with a corresponding cholesterol level measured on the same blood sample. We can see that there is a slight tendency for one measure to depend on the other, a fact that would not have been clear if we had simply listed each cholesterol level together with the corresponding triglyceride level.

The final graph that we shall mention here is the **tree diagram**. This is often used to help in making decisions, in which case it is called a **decision tree**. A tree diagram displays in temporal sequence possible types of actions or outcomes. Figure 3.8 gives a very simple example; it indicates the possible outcomes and their relative frequencies following myocardial infarction. This kind of display is often much more effective than a verbal description of the same infor-

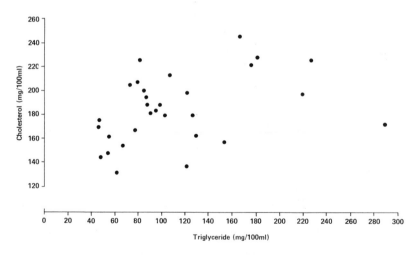

**FIGURE 3.7** Scatter plot of cholesterol versus triglyceride levels of 30 male medical students.

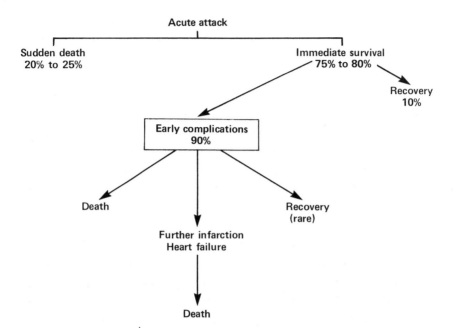

**FIGURE 3.8** Tree diagram indicating outcome of myocardial infarction. (Source: R.A. Cawson, A.W. McCracken and P.B. Marcus (1982) Pathologic mechanisms and human disease. C.V. Mosby Co.).

mation. Tree diagrams are also often helpful in solving problems, as we shall see in Chapter 4.

## PROPORTIONS AND RATES

In comparing the number of events occurring in two groups, raw numbers are difficult to interpret unless each group contains the same number of persons. We often compute proportions or percentages to facilitate such comparisons. Thus, if the purpose of a measure is to determine whether the inhabitants in one community have a greater or less frequent occurrence of tuberculosis than those in another, simple counts have obvious shortcomings. Community A may have more cases than Community B because its population is larger. To make a comparison, we need to know the proportionate number of cases in each community. Again, it may be necessary to specify the time at or during which the events of interest occur. Thus, if 500 new cases of tuberculosis were observed in a city of 2,000,000 persons in 1973, we say that 0.025% of the population in this city developed tuberculosis in 1973.

Sometimes it is more convenient to express proportions multiplied by some number other than 100 (which results in a percentage). Thus, the new cases of tuberculosis in a city for the year 1973 might be expressed as 500 cases per 2,000,000 persons (the actual population of the city), 0.025 per hundred (percent), 0.25 per thousand, or 250 per million. We see that three components are required for expressions of this type:

1. The number of individuals in whom the disease, abnormality, or other characteristic occurrence is observed (the numerator).
2. The number of individuals in the population among whom the characteristic occurrence is ascertained (the denominator).
3. A specified period of time during which the disease, abnormality, or characteristic occurrence is observed.

The numerator and the denominator should be similarly restricted; if the numerator represents a count of persons who have a characteristic in a particular age-race-sex group, then the denominator should also pertain to that same age-race-sex group. When the denominator is restricted solely to those persons who are capable of having or contracting a disease, it is sometimes referred to as a population at risk. For example, a hospital may express its maternal mortality as the number of maternal deaths per thousand deliveries. The women who delivered make up the population at risk for maternal deaths. Similarly, case fatality is the number of deaths due to a disease per so many patients with the disease; here the individuals with the disease constitute the population.

All such expressions are just conversions of counts into proportions, or fractions of a group, in order to summarize data so that comparisons can be made among groups. They are commonly called **rates**, though strictly speaking a rate is a measure of the rapidity of change of a phenomenon, usually per unit time. Expressions such as "maternal death rate" and "case fatality rate" are often used to describe these proportions even when no concept of a rate per unit time is involved. One of the main concerns of epidemiology is to find and enumerate appropriate denominators for describing and comparing groups in a meaningful and useful way. Two other commonly seen but often confused measures of disease frequency used in epidemiology are prevalence and incidence.

The **prevalence** of a disease is the number of cases (of that disease) at a given point in time. Prevalence is usually measured as the ratio of the number of cases at a given point in time to the number of people in the population of interest at that point in time.

The **incidence** of a disease is the number of new cases that occur during a specified amount of time. To adjust for the size of the population being observed, incidence is usually measured as the ratio of the number of new cases occurring during a period to the number of people initially at risk of developing the disease. Thus, if the number of cases of coronary heart disease events (such as angina, myocardial infarction, or sudden death) occurring in a population of 742 men during a 12-year period is observed to be 88, the incidence rate would be computed as follows:

$$I = \frac{88}{742} \times 1,000 = 119 .$$

Thus, the incidence is 119 coronary-heart-disease events per 1,000 persons initially at risk, during the 12-year period. If this were expressed **per year,** it would be a true rate.

Figure 3.9 demonstrates the difference between incidence and prevalence. Assume that each line segment represents a case of dis-

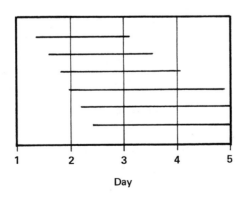

**FIGURE 3.9** Six cases of a disease represented over time by line segments.

ease from time of onset (beginning of the line segment) until the disease has run its course (end of the line segment). Moreover, assume that 1,000 persons are at risk on any given day. The incidence for day 1 is 4 cases per 1,000 persons (4 new line segments) and for day 2 it is 2 cases per 1,000 persons (2 new line segments). The prevalence at the end of day 1 is 4 per 1,000 (4 line segments exist), and at the end of day 2 it is 6 (6 line segments exist). It should be obvious that two diseases can have identical incidence, and yet one would have a much higher prevalence if its duration (time from onset until the disease has run its course) is much larger.

Incidence measures rate of development of disease. It is therefore a measure of risk of disease and is useful in studying possible reasons (or causes) for disease developing. We often study the incidence in different groups of people and then try to determine the reasons why it may be higher in one of the groups.

Prevalence measures the amount of disease in a population at a given point in time. Because prevalence is a function of the duration of a disease, it is of more use in planning health-care services for that disease.

## RELATIVE MEASURES OF DISEASE FREQUENCY

Several methods have been developed for measuring the relative amount of disease occurring in different populations. For example, we might wish to measure the amount of disease occurring in a group exposed to some environmental condition, such as cigarette smoking, relative to that in a group not exposed to that condition. One measure used for this purpose is the **relative risk**, which is defined as

$$RR = \frac{\text{incidence of disease in exposed group}}{\text{incidence of disease in unexposed group}}.$$

For example, if the incidence of a particular disease in a group exposed to some condition is 30 per 100,000 per year, compared with an incidence of 10 per 100,000 per year in a group unexposed to the condition, then the relative risk (exposed versus unexposed) is

$$RR = \frac{30 \text{ per } 100,000 \text{ per year}}{10 \text{ per } 100,000 \text{ per year}} = 3.$$

The phrase "exposed to a condition" is used in a very general sense. Thus one can talk of the relative risk of ankylosing spondylitis to a person possessing the HLA antigen B27, versus not possessing that antigen, though of course the antigen is inherited from a parent rather than acquired from some kind of exposure.

Another relative measure of disease occurrence is the **odds ratio.** The odds in favor of a particular event are defined as the frequency with which the event occurs divided by the frequency with which it does not occur. For a disease with an incidence 30 per 100,000 per year, for example, the odds in favor of the disease are $30/99,970$. The odds ratio is then defined as

$$\text{OR} = \frac{\text{odds in favor of disease in exposed group}}{\text{odds in favor of disease in unexposed group}} .$$

Thus, if the incidences are 30 per 100,000 and 10 per 100,000 as above, the odds ratio for exposed versus unexposed, is

$$\text{OR} = \frac{30}{99,970} \bigg/ \frac{10}{99,990} = 3.00006 .$$

You can see from this example that for rare diseases the odds ratio closely approximates the relative risk. The attractive feature of the odds ratio is that it can be estimated without actually knowing the incidences. This is often done in case-control studies, which were described in Chapter 2. Suppose, for example, it is found that 300 out of 1,000 cases of a disease had previous exposure to a particular condition, whereas only 100 out of 1,000 controls were similarly exposed. These data tell us nothing about the incidences of the disease among exposed and unexposed persons, but they do allow us to calculate the odds ratio, which in this case is

$$\frac{300}{700} \bigg/ \frac{100}{900} \cong 4.$$

The last relative measure of disease frequency we shall discuss is the **attributable risk,** defined as the incidence of disease is an exposed group minus the incidence of disease in an unexposed group. Thus, in the previous example, the attributable risk is

$$\text{AR} = 30 - 10 = 20 \text{ per } 100,000 \text{ per year.}$$

An excess of 20 cases per 100,000 per year can be attributed to exposure to the particular condition. Sometimes we express attributable risk as a percentage of the incidence of disease in the unexposed group. In the above example, we would have

$$AR\% = \frac{30 - 10}{10} \times 100 = 200\% \ .$$

In this case we could say there is a 200% excess risk of disease in the exposed group.

## SENSITIVITY, SPECIFICITY, AND PREDICTIVE VALUES

We now define some terms that are often used to measure the effectiveness of a test procedure, such as a laboratory test to help diagnose a disease. We shall illustrate these terms using the following hypothetical population of 10,000 persons classified on the basis of disease status and their reaction to the test:

| Disease status | Test Result | | Total |
| --- | --- | --- | --- |
| | Negative | Positive | |
| Absent | 8,820 | 980 | 9,800 |
| Present | 20 | 180 | 200 |
| Total | 8,840 | 1,160 | 10,000 |

Note first that the prevalence of the disease in the population is $200/10,000$, or 2%.

The **sensitivity** of the test measures how well it detects disease; it is the proportion of those with the disease who give a positive result. In the example the sensitivity is $180/200 = 0.9$.

The **specificity** of the test measures how well it detects absence of disease; it is the proportion of those without the disease who give a negative result. In the example, the specificity is $8,820/9,800 = 0.9$.

Whenever sensitivity and specificity are equal, they represent the proportion of the population that is correctly classified by the test. Thus, in our example, 90% of the total population is correctly classified by the test. This does not mean, however, that 90% of those who give a positive result have the disease. In order to know how to interpret a particular test result, we need to know the **predictive values** of the test, which are defined as the proportion of those positive who have the disease, and the proportion of those negative who do not have the disease. For our example, these values are $^{180}/_{1,160} = 0.155$, and $^{8,820}/_{8,840} = 0.998$, respectively. Especially in the case of a rare disease, a high specificity and high sensitivity are not sufficient to ensure that a large proportion of the positives actually have the disease.

## MEASURES OF CENTRAL TENDENCY

Measures of central tendency, or measures of location, tell us where on our scale of measuremeñt a set of values tends to lie. All the values in Table 3.1, for example, lie between 45 mg/100 ml and 287 mg/100 ml, and we need our measure of central tendency to be somewhere between these two values. If our values had been in mg/L, on the other hand, we should want our measure of central tendency to be ten times as large. We shall discuss three measures of central tendency: the mean, the median, and the mode. They all have the property (when used to describe continuous data) that if every value in our dataset is multiplied by a constant number, then the measure of central tendency is multiplied by the same number. Similarly, if a constant is added to every value, then the measure of central tendency is increased by that same amount.

The **mean** of a set of numbers is just their numerical average. You know, for example, that to compute your mean score for four test grades you add the grades and divide by four. If your grades were 94, 95, 97, and 98, your mean score would be $(94 + 95 + 97 + 98)/4 = 384/4 = 96$.

One of the disadvantages of the mean as a summary statistic is that it is sensitive to unusual values. The mean of the numbers 16, 18, 20, 22 and 24 is 20, and indeed 20 appears to be typical of these numbers. The mean of the numbers 1, 2, 3, 4 and 90 is also 20, but 20 is not typical of these numbers because of the one unusual value. Another disadvantage of the mean is that strictly speaking it should be used only for data measured on an interval scale, since implicit in its use is the assumption that the units of the scale are all of equal value. The difference between 50 and 51 mg/100 ml of triglyceride is in fact the same as the difference between 250 and 251 mg/100 ml of triglyceride, (i.e., 1 mg/100 ml). Because of this, it is meaningful to say that the mean of the 30 values in Table 3.1 is 111.2 mg/100 ml. But if the 30 students had been scored on an eleven-point scale, 0 through 10 (whether for triglyceride level or anything else), the mean score would be strictly appropriate only if each of the ten intervals, 0 to 1, 1 to 2, etc., were equal in value. Nevertheless, the mean is the most frequently used descriptive statistic because, as we shall see later, it has statistical properties that make it very advantageous if no unusual values are present.

The **median** is the middlemost value in a set of ranked data. Thus the median of the numbers 16, 18, 20, 22, and 24 is 20. The median of the numbers 1, 2, 3, 4, and 90 is 3. In both sets of numbers the median is typical of most of the data, so the median has the advantage of not being sensitive to unusual values. If the set of data contains an even number of values, then the median lies between the two middlemost values, and we usually just take their average. Thus the median of the data in Table 3.1 lies between 90 and 93 mg/100 ml, and we would usually say the median is 91.5 mg/100 ml.

Recall that a percentile is the value of a trait at or below which the corresponding percentage of a dataset lies. If your grade on an examination is at the 90th percentile, then 90% of those taking the examination obtained the same or a lower grade. The median is thus the 50th percentile—the point at or below which 50% of the data points lie. The median is a proper measure of central tendency for data measured either on an interval or on an ordinal scale, but cannot be used for nominal data.

The **mode** is defined as the most frequently occurring value in a set of data. Thus, for the data 18, 19, 21, 21, 22, the value 21 occurs twice, whereas all other values occur only once, and so 21 is the mode. In the case of continuous data, the mode is related to the concept of a peak in the frequency distribution. If there is only one peak, the distribution is said to be unimodal; if there are two peaks, it is said to be bimodal, etc. Thus the distribution depicted in Figure 3.1 is unimodal, and the mode is clearly between 180 and 190 mg/100 ml.

An advantage of the mode is that it can be used for nominal data: the modal category is simply the category that occurs most frequently. But it is often difficult to use for a small sample of continuous data. What, for example, is the mode of the data in Table 3.1? Each value occurs exactly once, so shall we say there is no mode? The data can be grouped as in Figure 3.2, and then it appears that the 70–90 mg/100 ml category is the most frequent. But with this grouping we also see peaks (and hence modes) at 150–170, 210–230, and 270–290 mg/100 ml. For this reason the mode is less frequently used as a measure of central tendency in the case of continuous data.

## MEASURES OF SPREAD OR VARIABILITY

Suppose you score 80 % on an examination and the average for the class is 87 %. Suppose you are also told that the grades ranged from 79 % to 95 %. Obviously you would feel much better had you been told that the spread was from 71 % to 99 %. The point here is that it is often not sufficient to know the mean of a set of data but, rather, it is of interest to know the mean together with some measure of spread or variability.

The **range** is the largest value minus the smallest value. It provides a simple measure of variability but is very sensitive to one or two extreme values. The range of the data in Table 3.1 is $287 - 45 = 242$ mg/100 ml, but it would be only 173 mg/100 ml if the two largest values were missing. Percentile ranges are less sensitive and provide a useful measure of dispersion in data. For example, the 90th percentile minus the 10th percentile, or the 75th percentile minus the 25th percentile can be used. The latter is called the **interquartile range**. For the data in Table 3.1 the interquartile range is $124 - 67 = 57$ mg/100 ml. (For 30 values we cannot obtain the 75th and 25th percentiles accurately, so we take the next lowest percentiles: 124 is the 22nd out of 30 values, or 73rd percentile, and 67 is the 7th out of 30, or 23rd percentile.) If the two largest values were missing from the table, the interquartile range would be $123 - 67 = 56$ mg/100 ml, i.e., almost the same as for all 30 values.

The **variance** or its square root, the **standard deviation**, is perhaps the most frequently used measure of variability. The variance is basically the average squared deviation from the mean. To compute the variance of a set of data, we:

1. Subtract the mean from each value to get a "deviation" from the mean.
2. Square each deviation from the mean.
3. Sum the squares of the deviations from the mean.
4. Divide the sum of squares by one less than the number of values in the set of data.

Thus, for the numbers 18, 19, 20, 21, and 22, we find that the mean is $(18 + 19 + 20 + 21 + 22)/5 = 20$, and the variance is computed as follows:

1. Subtract the mean from each value to get a deviation from the mean, which we shall call d:

$$
\begin{array}{c}
d \\
\hline
18 - 20 = -2 \\
19 - 20 = -1 \\
20 - 20 = \phantom{+}0 \\
21 - 20 = +1 \\
22 - 20 = +2
\end{array}
$$

2. Square each deviation, d, to get squares of deviations, $d^2$:

| d | $d^2$ |
|---|---|
| -2 | 4 |
| -1 | 1 |
| 0 | 0 |
| +1 | 1 |
| +2 | 4 |

3. Sum the squares of the deviations:

$$4 + 1 + 0 + 1 + 4 = 10.$$

4. Divide the sum of squares by one less than the number of values in the set of data.

$$\text{Variance} = s^2 = \frac{10}{5-1} = \frac{10}{4} = 2.5 \ .$$

The standard deviation is just the square root of the variance, that is, in this example,

$$\text{Standard deviation} = s = \sqrt{2.5} = 1.6 \; .$$

Notice that the variance is expressed in squared units, whereas the standard deviation gives results in terms of the original units. If, for example, the original units for the above data were years (e.g., years of age), then $s^2$ would be 2.5 (years)$^2$, and s would be 1.6 years.

As a second example, suppose the numbers were 1, 2, 3, 4, and 90. Again, the average is $(1 + 2 + 3 + 4 + 90)/5 = 20$, but the data are quite different from those in the previous example. Here,

$$
\begin{aligned}
s^2 &= [(1 - 20)^2 + (2 - 20)^2 + (3 - 20)^2 + (4 - 20)^2 + (90 - 20)^2]/4 \\
&= [(-19)^2 + (-18)^2 + (-17)^2 + (-16)^2 + (70)^2]/4 \\
&= 6130/4 = 1532.5 \; .
\end{aligned}
$$

The standard deviation is $s = \sqrt{1532.5} = 39.15$ . Thus, you can see that the variance and the standard deviation are larger for a set of data that is obviously more variable.

Two questions you may have concerning the variance and the standard deviation are: Why do we square the deviations, and why do we divide by one less than the number of values in the set of data being considered? Look back at Step 2 and see what would happen if we didn't square the deviations, but simply added up the unsquared deviations, d. Because of the way in which the mean is defined, the deviations always add up to zero! Squaring is a simple device to stop this happening. When we average the squared deviations, however, we divide by one less than the number of values in the data set. The reason for this is that it leads to an unbiased estimate, a concept we shall explain more fully in Chapter 6. For the moment, just note that if the data set consisted of an infinite number of values (which is conceptually possible for a whole population), it would make no difference whether or not we subtracted one from the divisor.

The last measure of spread we shall discuss is the **coefficient of variation**. This is the standard deviation expressed as a proportion or percentage of the mean. It is a dimensionless measure and, as such, it is a useful descriptive index for comparing the relative variability in two sets of values. Suppose, for example, that we wished to compare the variability of birthweight with the variability of adult weight. Clearly, on an absolute scale, birthweights must vary much less than adult weights simply because they are necessarily limited to being much smaller. As a more extreme example, suppose we wished to compare the variability in weights of ants and elephants! In such a situation it makes more sense to express variability on a relative scale. Thus, we can make a meaningful comparison of the variability in two sets of numbers with different means by observing the difference between their coefficients of variation. As an example, suppose the mean of a set of cholesterol levels is 219 mg/100 ml and the standard deviation is 14.3 mg/100 ml. The coefficient of variation, as a percentage, is then

$$CV\% = \frac{\text{standard deviation}}{\text{mean}} \times 100$$

$$= \frac{14.3 \text{ mg}/100 \text{ ml}}{219 \text{ mg}/100 \text{ ml}} \times 100$$

$$= 6.5 \ .$$

This could then be compared with, for example, the coefficient of variation of triglyceride levels.

## MEASURES OF SHAPE

There are many other descriptive statistics, some of which will be mentioned in later chapters of this book. We shall conclude this chapter with the names of a few statistics that describe the shape of distributions. (Formulae for calculating these statistics, as well as others, are presented in Appendix 1).

The **coefficient of skewness** is a measure of symmetry. A symmetric distribution has a coefficient of skewness that is zero. As illustrated in Figure 3.10, a distribution that has a tail to the left has a negative coefficient of skewness, and is said to be negatively skewed; one that has a tail to the right has a positive coefficient of skewness, and is said to be positively skewed. Note that in a symmetric unimodal distribution the mean, the median, and the mode are all equal. In a unimodal asymmetric distribution the median always lies between the mean and the mode. The serum triglyceride values in Table 3.2 have

mean | mode
median

Negatively skewed

mean
mode
median

Symmetric

mode | mean
median

Positively skewed

**FIGURE 3.10** Examples of negatively skewed, symmetric, and positively skewed distributions.

a positive coefficient of skewness, as can be seen in the histogram in Figure 3.4.

The **coefficient of kurtosis** measures the peakedness of a distribution. In Chapter 6 we shall discuss a very important distribution, called the normal distribution, for which the coefficient of kurtosis is 3. A distribution with a larger coefficient than this is leptokurtic ("lepto" means slender), and one with a coefficient smaller than this is platykurtic ("platy" means flat or broad). Kurtosis, or peakedness, is also often measured by the standardized "fourth cumulant" (denoted $g_2$), which is the coefficient of kurtosis minus three; on this scale, the normal distribution has zero kurtosis. Different degrees of kurtosis are illustrated in Figure 3.11.

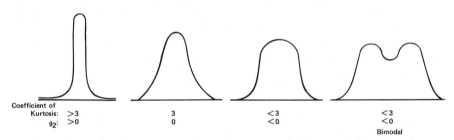

Coefficient of
Kurtosis:  >3
$g_2$:  >0

3
0

<3
<0

<3
<0
Bimodal

**FIGURE 3.11** Examples of distributions with coefficient of kurtosis greater than 3 ($g_2>0$), equal to three ($g_2=0$, as for a normal distribution), and less than three ($g_2<0$).

## SUMMARY

1. Continuous data arise only from quantitative traits, whereas categorical or discrete data arise either from quantitative or from qualitative traits. Continuous data are measured on an interval scale,

categorical data on either an ordinal (that can be ordered) or a nominal (name only) scale.

2. Descriptive statistics, tables, and graphs summarize the essential characteristics of a set of data.

3. A table should be easy to read. The title should indicate what is being tabulated, with the units of measure.

4. Bar graphs are used for discrete data, histograms and frequency polygons for continuous data. A cumulative plot has the advantage that every data-point can be represented in it. A scatter plot or scatter diagram illustrates the relationship between two measures. A tree diagram displays a sequence of actions and/or results.

5. Proportions and rates allow one to compare counts when denominators are appropriately chosen. The term "rate" properly indicates a measure of rapidity of change, but is often used to indicate a proportion multiplied by some number other than 100. Prevalence is the number or proportion of cases present at a particular time, incidence is the number or proportion of new cases occurring in a specified period.

6. Relative risk is the incidence of disease in a group exposed to a particular condition, divided by the incidence in a group not so exposed. The odds ratio is the ratio of the odds in favor of a disease in an exposed group to the odds in an unexposed group. In the case of a rare disease, the relative risk and the odds ratio are almost equal. Attributable risk is the incidence of a disease in a group with a particular condition minus the incidence in a group without the condition, often expressed as a percentage of the latter.

7. The sensitivity of a test is the proportion of those with the disease who give a positive result. The specificity of a test is the proportion of those without the disease who give a negative result. Even though these are both high, in the case of a rare disease it is quite possible for the test to have a low predictive value. The predictive values are defined as the proportion of the positives who have the disease, and the proportion of the negatives who do not have the disease.

8. Three measures of central tendency, or location, are the mean (arithmetic average), the median (50th percentile), and the mode (one or more peak values). All three are equal in a unimodal symmetric distribution. In a unimodal asymmetric distribution, the median lies between the mean and the mode.

9. Three measures of spread, or variability, are the range (largest value minus the smallest value), the interquartile range (75th percentile minus the 25th percentile), and the standard deviation (square root of the variance). The variance is basically the average squared deviation from the mean, but the divisor used to obtain this average is one less than the number of values being averaged. The variance is expressed in squared units. The coefficient of variation, which is dimensionless, is the standard deviation divided by the mean (and multiplied by 100 if it is expressed as a percentage).

10. An asymmetric distribution may be positively skewed (tail to the right) or negatively skewed (tail to the left). A distribution may be leptokurtic (peaked) or platykurtic (flat-topped or multimodal).

### FURTHER READING

Stevens, S.S. (1946) On the Theory of Scales of Measurement. Science 103:677–680. (This article defines in more detail four hierarchical categories of scales for measurements—nominal, ordinal, interval, and ratio.)

Elandt-Johnson, R.C. (1975) Definition of Rates: Some Remarks on Their Use and Misuse. American Journal of Epidemiology 102:267–271. (This gives very precise definitions of ratios, proportions, and rates; a complete understanding of this paper requires some mathematical sophistication.)

Wainer, H. (1984) How to Display Data Badly. The American Statistician 38:137–147. (Though pointed in the wrong direction, this is a serious article. It illustrates the 12 most powerful methods—the dirty dozen—of misusing graphics.)

### PROBLEMS

1. The following cumulative plot was derived from the pulse rates of 1,000 students:

Pulse rate in beats per minute

Which of the following is false?
A. The range of the distribution is 60–100 beats/minute.
B. The mode of the distribution is 100 beats/minute.
C. The median of the distribution is 77 beats/minute.
D. 94% of the values are less than 90 beats/minute.
E. 94% of the values are greater than 65 beats/minute.

2. The following is a histogram for the pulse rates of 1,000 students:

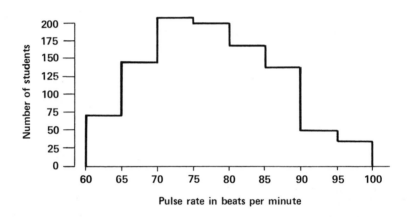

Pulse rate in beats per minute

Which of the following is between 70 and 75 beats/minute?
A. The mode of the distribution.
B. The median of the distribution.
C. The mean of the distribution.
D. The range of the distribution.
E. None of the above.

3. A nominal scale is used for
A. all categorical data.
B. discrete data with categories that do not follow a natural sequence.
C. continuous data that follow a natural sequence.
D. discrete data with categories that follow a natural sequence.
E. quantitative data.

4. The following are Average Annual Incidences per 1,000,000 for Testicular Cancers, New Orleans, 1974 to 1977.

| Age | White | Black | Odds Ratio |
|-----|-------|-------|-----------|
| 15–19 | 29.4 | 13.4 | 2.2 |
| 20–29 | 113.6 | 9.5 | 12.0 |
| 30–39 | 91.0 | 49.8 | 1.8 |
| 40–49 | 75.5 | 0.0 | — |
| 50–59 | 50.2 | 22.2 | 2.3 |
| 60–69 | 0.0 | 0.0 | — |
| 70+ | 38.2 | 0.0 | — |

Based on these data, which of the following is true of New Orleans males?
  A. There is no difference in the risk of developing testicular cancer for blacks and whites.
  B. The odds of developing testicular cancer are greater in blacks than in whites.
  C. The racial difference in risk of developing testicular cancer cannot be determined from these data.
  D. The risk of developing testicular cancer is greater in whites than in blacks in virtually every age group.

5. Consider the following sets of cholesterol levels in mg/100 ml.
            Set 1. 200, 210, 190, 220, 195
            Set 2. 210, 170, 180, 235, 240
The standard deviation of Set 1 is
  A. the same as that of Set 2.
  B. less than that of Set 2.
  C. greater than that of Set 2.
  D. equal to the mean for Set 2.
  E. indeterminable from these data.

6. A large study of bladder cancer and cigarette smoking produced the following data:

|  | Incidence of bladder cancer (per 100,000 males per year) |
|-----|-----|
| Cigarette smokers | 48.0 |
| Nonsmokers | 25.4 |

The relative risk of developing bladder cancer for male cigarette smokers compared with male nonsmokers is

A. $48.0/25.4 = 1.89$
B. unknown
C. $48.0 - 25.4 = 22.6$
D. $48.0$
E. $\dfrac{48.0 - 25.4}{48.0}$

7. Refer to the diagram below. Each horizontal line in the diagram indicates the month of onset and the month of termination for one of 24 episodes of disease. Assume an exposed population of 1,000 individuals in each month.

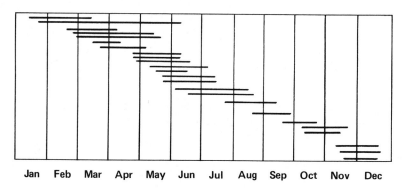

| Jan | Feb | Mar | Apr | May | Jun | Jul | Aug | Sep | Oct | Nov | Dec |

(1) The incidence for this disease during April was
A. 2 per 1,000.
B. 3 per 1,000.
C. 6 per 1,000.
D. 7 per 1,000.
E. 9 per 1,000.

(2) The prevalence on March 31 was
A. 2 per 1,000.
B. 3 per 1,000.
C. 6 per 1,000.
D. 7 per 1,000.
E. 9 per 1,000.

8. One useful summary of a set of data is provided by the mean and standard deviation. Which of the following is true?

A. The mean is the middle value (50th percentile) and the standard deviation is the difference between the 90th and the 10th percentiles.
B. The mean is the arithmetic average and the standard deviation measures the extent to which observations vary about or are different from the mean.
C. The mean is the most frequently occurring observation and the standard deviation measures the length of a deviation.
D. The mean is half the sum of the largest and smallest value and the standard deviation is the difference between the largest and smallest observations.
E. None of the above.

9. Which of the following are measures of spread?
(1) Mode.
(2) Range.
(3) Skewness.
(4) Standard deviation.
   A. 1, 2, and 3
   B. 1 and 3
   C. 2 and 4
   D. Only 4
   E. 1, 2, 3 and 4

10. Both the specificity and sensitivity of a diagnostic test for a particular disease are 0.99. It follows that
(1) a person with the disease has a 99% chance of being positive for the test.
(2) a person without the disease has a 99% chance of being negative for the test.
(3) a person has a 99% chance of being correctly classified by the test.
(4) a person who is positive for the test has a 99% chance of having the disease.
   A. 1, 2, and 3
   B. 1 and 3
   C. 2 and 4
   D. Only 4
   E. 1, 2, 3 and 4

CHAPTER 4

# The Laws of Probability

**Key Concepts:** three definitions of probability; the probability of either of two events, the joint probability of two events, conditional probability, independent events; Bayes' theorem; likelihood ratio.

## DEFINITION OF PROBABILITY

Although the meaning of a statement such as, "The probability of rain today is fifty percent," may seem fairly obvious, it is not easy to give an exact definition of the term "probability." In fact, three different definitions have been proposed, each focusing on a different aspect of the concept. Since probability plays such a fundamental role in biostatistics, we start by reviewing all three of these definitions before stating the mathematical laws that govern its manipulation. The mathematicians who originally studied probability were motivated by gambling, and so used games of chance (e.g., cards and dice) in their studies. It will be convenient for us to use similar examples initially.

**The Classical Definition of Probability** can be stated as follows: Given a set of equally likely possible outcomes, the probability P of the event A [which we write P(A)] is the number of outcomes that are "favorable to" A divided by the total number of possible outcomes:

$$P(A) = \frac{\text{Number of outcomes favorable to A}}{\text{Total number of possible outcomes}}.$$

This definition will become clearer when we illustrate it with some examples. Suppose we pick a single card at random from a well-shuffled regular deck of 52 cards comprising four suits, with 13 cards in each suit. What is the probability that the card is an ace? We let A

**60**

be the event "the card is an ace." Each card represents a possible outcome, and so the total number of possible outcomes is 52; the number of outcomes that are "favorable to" A is four, since there are four aces in the deck. Therefore the probability is

$$P(A) = \frac{4}{52} = \frac{1}{13}.$$

As another example, what is the probability of obtaining two sixes when two normal six-sided dice are rolled? In this case let A be the event "obtaining two sixes." Each of the two dice can come up one, two, three, four, five, or six, and so the total number of possible outcomes is 36 (6 × 6). Only one of these outcomes (six and six) is "favorable to" A, and so the probability is $P(A) = 1/36$.

This definition of probability is precise, and appears to make sense. Unfortunately, however, it contains a logical flaw. Note that the definition includes the words "equally likely," which is another way of saying "equally probable." In other words, **probability has been defined in terms of probability!** Despite this difficulty, our intuition tells us that when a card is taken at random from a well-shuffled deck, or when normal dice are rolled, there is physical justification for the notion that all possible outcomes are "equally likely." When there is no such physical justification (as we shall see in a later example), the classical definition can lead us astray if we are not careful.

**The Frequency Definition of Probability** supposes that we can perform the same trial or experiment many, many times. As the number of trials tends to infinity (i.e., to any very large number, represented mathematically by the symbol ∞), the proportion of trials in which the event A occurs tends to a fixed limit. We then define the probability of the event A as this limiting proportion. Thus, to answer the question "What is the probability that the card is an ace?", we suppose that many trials are performed, in each of which a card is drawn at random from a well-shuffled deck of 52 cards. After each trial we record in what proportion of the cards drawn so far an ace has been drawn, and we find that, as the number of trials increases indefinitely, this proportion tends to the limiting value of $1/13$. This concept is illustrated in Figure 4.1, which represents a set of trials in which an ace was drawn at the 10th, 20th, 40th, 54th, 66th, and 80th trials.

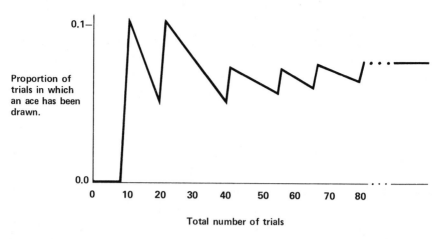

**FIGURE 4.1** Example of probability defined as the limiting proportion of trials, as the number of trials tends to infinity, in which a particular event occurs.

Since one can never actually perform an infinite number of trials, this definition of probability is mathematically unsatisfying. Nevertheless, it is the best way of interpreting probability to a patient in practical situations. The statement "There is a fifty-percent probability of rain today" can be interpreted to mean that on just half of many days on which such a statement is made it will rain. Provided that this is true, the probability statement is valid.

In medical practice, however, quoting valid probabilities may not be sufficient. When a pregnant woman is counseled concerning the probability that her child will have a particular disease, the probability that is quoted must be both valid and **relevant**. Suppose, for example, a 25-year-old woman has a child with Down's syndrome (trisomy 21), and, on becoming pregnant again, seeks counseling. Her physician checks published tables and finds out that the probability of a 25-year-old woman having a Down's syndrome child is about 1 in 400. If she is counseled that the probability of her second child having Down's syndrome is only about 1 in 400, she will have been quoted a valid probability: only 1 in 400 such women coming for counseling will bear a child with Down's syndrome. But this probability will not be relevant if the first child has Down's syndrome because of a translocation in the mother, i.e., because the mother has a long arm of chromosome 21 attached to another of her chromosomes. If this is the case, the risk to the unborn child is about 1 in 6. A physician who failed to recommend the additional testing required to arrive at such a relevant probability for a specific patient could face a malpractice suit, even though the quoted probability, on the basis of all the information available, was perfectly valid.

**The Mathematical (Axiomatic) Definition of Probability** avoids the disadvantages of the two other definitions, and is the definition used by mathematicians. A simplified version of this definition, which, although incomplete, retains its main features, is as follows. A set of probabilities is any set of numbers for which: (1) each number is **positive** (greater than or equal to zero), and (2) the sum of the numbers is unity (one). This definition, unlike the other two, gives no feeling for the practical meaning of probability. It does, however, describe the essential characteristics of probability: there is a set of possible outcomes, each associated with a positive probability of occurring, and at least one of these outcomes must occur. We now turn to some fundamental laws of probability, which do not depend on which definition is taken.

## THE PROBABILITY OF EITHER OF TWO EVENTS: A OR B

If A and B are two events, what is the probability of either A or B occurring? Let us consider a simple example. A regular deck of cards is shuffled well and a card is drawn. What is the probability that it is either an ace or a king? The deck contains eight cards that are either aces or kings, and so the answer is $8/52 = 2/13$. Now notice that the two events "ace" and "king" are mutually exclusive, in that when we draw a card from the deck it cannot be both an ace and a king. If A and B are mutually exclusive events, then

$$P(A \text{ or } B) = P(A) + P(B).$$

Thus, in our example, we have

$$P(\text{ace or king}) = P(\text{ace}) + P(\text{king})$$
$$= \frac{1}{13} + \frac{1}{13}$$
$$= \frac{2}{13}.$$

Now suppose the question had been: "What is the probability that the card is an ace or a heart?" In this case the events "ace" and "heart" are **not** mutually exclusive, because the same card could be both an ace and a heart; therefore we cannot use the formula given above. How many cards in the deck are either aces or hearts? There are 4 aces and 13 hearts, but a total of only 16 cards that are either an ace or a heart: the ace of clubs, the ace of diamonds, the ace of spades and the 13 hearts. Notice that if we carelessly write "4 aces + 13 hearts = 17 cards," the ace of hearts has been counted twice—once as an ace and once as a heart. In other words, the number of cards that are either aces or hearts is the number of aces, plus the number of hearts, minus the number of cards that are both aces and hearts (one, in this example). Analogously, dividing each of these numbers by 52 (the total number of cards in the deck), we have:

$$P(\text{ace or heart}) = P(\text{ace}) + P(\text{heart}) - P(\text{ace and heart})$$
$$= \frac{4}{52} + \frac{13}{52} - \frac{1}{52}$$
$$= \frac{16}{52} = \frac{4}{13}.$$

The general rule for any two events A and B is:

$$P(A \text{ or } B) = P(A) + P(B) - P(A \text{ and } B).$$

This rule is **general,** by which we mean it is **always** true. (A layman uses the word "generally" to mean "usually." In mathematics a **general** result is one that is always true, and the word "generally" means "always.")

If the event A is "the card is a king" and the event B is "the card is an ace," we have

$$P(A \text{ or } B) = P(A) + P(B) - P(A \text{ and } B)$$
$$= \frac{1}{13} + \frac{1}{13} - 0$$
$$= \frac{2}{13}, \text{ as before.}$$

In this case $P(A \text{ and } B) = 0$ because a card cannot be both a king and an ace. In the special case in which A and B are mutually exclusive events, $P(A \text{ and } B) = 0$.

# THE JOINT PROBABILITY OF TWO EVENTS: A AND B

We have just seen that the probability that both events A and B occur is written P(A and B); this is also sometimes abbreviated to P(AB) or P(A,B). It is often called the **joint** probability of A and B. If A and B are mutually exclusive (i.e. they cannot both occur—if a single card is drawn, for example, it cannot be both an ace and a king), then their joint probability, P(A and B), is zero. What can we say about P(A and B) in general? One answer to this question is implicit in the general formula for P(A or B) just given. Rearranging this formula we find:

$$P(A \text{ and } B) = P(A) + P(B) - P(A \text{ or } B) .$$

A more useful expression, however, uses the notion of **conditional** probability: the probability of an event occurring given that another event has already occurred. We write the **conditional probability of B occurring given that A has occurred** as P(B|A). Read the vertical line as "given," so that P(B|A) is the "probability of B given A" (i.e., given that A has already occurred). Sensitivity and specificity are examples of conditional probabilities. We have:

Sensitivity = P(test positive|disease present)
Specificity = P(test negative|disease absent).

Using this concept of conditional probability, we have the following general rule for the joint probability of A and B:

$$P(A \text{ and } B) = P(A) \, P(B|A) .$$

Since it is arbitrary which event we call A and which B, note we could equally well have written:

$$P(A \text{ and } B) = P(B) \, P(A|B) .$$

Using this formula, what is the probability, on drawing a single card from a well-shuffled deck, that it is both an ace and a heart (i.e., that it is the ace of hearts)? Let A be the event that the card is an ace and B be the event that it is a heart. Then, from the formula:

$$P(\text{ace and heart}) = P(\text{ace}) \, P(\text{heart}|\text{ace})$$
$$= \frac{1}{13} \cdot \frac{1}{4} = \frac{1}{52} .$$

$P(\text{heart} \mid \text{ace}) = 1/4$, since the ace that we have picked can be any of the four suits, only one of which is hearts. Similarly, we could have found

$$P(\text{ace and heart}) = P(\text{heart})\, P(\text{ace} \mid \text{heart})$$
$$= \frac{1}{4} \cdot \frac{1}{13} = \frac{1}{52} .$$

Now notice that in this particular example we have

$$P(\text{heart} \mid \text{ace}) = P(\text{heart}) = \frac{1}{4}$$

$$\text{and } P(\text{ace} \mid \text{heart}) = P(\text{ace}) = \frac{1}{13} .$$

In other words, the probability of picking a heart is the same ($1/4$) whether or not an ace has been picked, and the probability of picking an ace is the same ($1/13$) whether or not a heart has been picked. These two events are therefore said to be **independent**.

Two events A and B are independent if $P(A) = P(A \mid B)$, or if $P(B) = P(B \mid A)$. From the general formula for $P(A \text{ and } B)$, it follows that if two events A and B are independent, then

$$P(A \text{ and } B) = P(A)\, P(B)$$

Conversely, two events A and B are independent if we know that

$$P(A \text{ and } B) = P(A)\, P(B) .$$

It is often intuitively obvious when two events are independent. Suppose we have two regular decks of cards and randomly draw one card from each. What is the probability that the card from the first deck is a king and the card from the second deck is an ace? The two draws are clearly independent, since the outcome of the first draw cannot in any way affect the outcome of the second draw. The probability is thus

$$\frac{1}{13} \cdot \frac{1}{13} = \frac{1}{169} .$$

But suppose we have only one deck of cards, from which we draw two cards consecutively. Now what is the probability that the first is a king and the second is an ace? Using the general formula for the joint probability of two events A and B, we have:

$$P(\text{1st is king and 2nd is ace})$$
$$= P(\text{1st is king}) \; P(\text{2nd is ace} \mid \text{1st is king})$$
$$= \frac{4}{52} \cdot \frac{4}{51} = \frac{4}{663} \; .$$

In this case the two draws are not independent. The probability that the second card is an ace depends on what the first card is (if the first card is an ace, for example, the probability that the second card is an ace becomes $3/51$).

## EXAMPLES OF INDEPENDENCE, NON-INDEPENDENCE, AND GENETIC COUNSELING

It is a common mistake to assume two events are independent when they are not. Suppose two diseases occur in a population and it is known that it is impossible for a person to have both diseases. There is a strong temptation to consider two such diseases to be occurring independently in the population, whereas in fact this is impossible. Can you see why this is so? [Hint: let the occurrence of one disease in a particular individual be the event A, and the occurrence of the other disease be event B; what do you know about the joint probability of A and B if: (1) they are independent, and (2) they are mutually exclusive? Can P(A and B) be equal to both P(A)P(B) and zero if P(A) and P(B) are both non-zero (we are told both diseases actually occur in the population)?] On the other hand, it is sometimes difficult for an individual to believe that truly independent events occur in the manner in which they do occur. The mother of a child with a genetic anomaly may be properly counseled that she has a 25 % probability of having a similarly affected child at each conception, and that all conceptions are independent. But she will be apt to disbelieve the truth of such a statement if (as will happen to one quarter of mothers in this predicament, assuming the counseling is valid and she has another child) her very next child is affected.

## NON-INDEPENDENCE DUE TO MULTIPLE ALLELISM

Among the white population, 44% have red blood cells that are agglutinated by an antiserum denoted anti-A, and 56% do not. Similarly, the red blood cells of 14% of the population are agglutinated by another antiserum denoted anti-B, and those of 86% are not. If these two traits are distributed independently in the population, what proportion would be expected to have red cells that are agglutinated by both anti-A and anti-B? Let A+ be the event that a person's red cells are agglutinated by anti-A, and B+ the event that they are agglutinated by anti-B. Thus $P(A+) = 0.44$, $P(B+) = 0.14$. If these two events are independent, we should expect

$$P(A+ \ B+) = P(A+) \ P(B+) = (0.44)(0.14) = 0.06 \ .$$

In reality, less than 4% of Caucasians fall into this category (i.e., have an AB blood type); therefore, the two traits are not distributed independently in the population. This finding was the first line of evidence used to argue that these two traits are due to multiple alleles at one locus (the ABO locus), rather than to segregation at two separate loci.

## NON-INDEPENDENCE DUE TO LINKAGE DISEQUILIBRIUM

The white cells of 31% of the Caucasian population react positively with HLA anti-A1, and those of 21% react positively with HLA anti-B8 (HLA denotes the human leukocyte antigen system). If these two traits are independent, we would expect the proportion of the population whose cells would react positively to both antisera to be $(0.31)(0.21) = 0.065$ (i.e., 6.5%). In fact, we find that 17% of the population are both A1 positive and B8 positive. In this case family studies have shown that two loci (A and B), very close together on chromosome 6, are involved. Non-independence between two loci that are close together is termed linkage disequilibrium.

These two examples illustrate the fact that more than one biological phenomenon can lead to lack of independence. In many cases in the literature, non-independence is established and then, on the basis of that evidence alone, a particular biological mechanism is incorrectly inferred.

## CONDITIONAL PROBABILITY IN GENETIC COUNSELING

We shall now consider a very simple genetic example that will help you learn how to manipulate conditional probabilities. First, recall the formula for the joint probability of events A and B, which can be written (backwards) as

$$P(A) \, P(B \mid A) = P(A \text{ and } B) \, .$$

Now divide both sides by P(A) [which may be done provided P(A) is not zero], and we find

$$P(B \mid A) = \frac{P(A \text{ and } B)}{P(A)} \, .$$

This gives us a formula for calculating the conditional probability of B given A, $P(B \mid A)$, if we know $P(A \text{ and } B)$ and $P(A)$. Similarly, if we know $P(A \text{ and } B)$ and $P(B)$, we can find

$$P(A \mid B) = \frac{P(A \text{ and } B)}{P(B)} \quad [\text{assuming } P(B) \text{ is not zero}] \, .$$

Now consider, as an example, hemophilia—a rare X-linked recessive disease in which there is a defect in the blood-clotting system. A carrier mother who marries an unaffected man transmits the disease to half her sons and to none of her daughters (half her daughters will be carriers, but none will be affected with the disease). What is the probability that she will bear an affected child? The child must be either a son or a daughter, and these are mutually exclusive events. Therefore

$$
\begin{aligned}
P(\text{affected child}) &= P(\text{affected son}) + P(\text{affected daughter}) \\
&= P(\text{son and affected}) + P(\text{daughter and} \\
&\quad \text{affected}) \\
&= P(\text{son})P(\text{affected} \mid \text{son}) + P(\text{daughter}) \\
&\quad \times P(\text{affected} \mid \text{daughter}).
\end{aligned}
$$

Assume a 1:1 sex ratio [i.e., $P(\text{son}) = P(\text{daughter}) = 1/2$], and use the fact that $P(\text{affected} \mid \text{son}) = 1/2$. Furthermore, in the absence of mutation (which has such a small probability that we shall ignore it), $P(\text{affected} \mid \text{daughter}) = 0$. We therefore have

$$P(\text{affected child}) = \frac{1}{2} \cdot \frac{1}{2} + \frac{1}{2} \cdot 0 = \frac{1}{4}.$$

Now suppose an amniocentesis is performed and thus the sex of the fetus determined. If the child is a daughter, the probability of her being affected is virtually zero. If, on the other hand, the child is a son, the probability of his being affected is one-half. Note that we can derive this probability by using the formula for conditional probability:

$$P(\text{affected} \mid \text{son}) = \frac{P(\text{son and affected})}{P(\text{son})}$$

$$= \frac{\dfrac{1}{4}}{\dfrac{1}{2}} = \frac{1}{2}.$$

Of course in this case you do not need to use the formula to obtain the correct answer, but you should nevertheless be sure to understand the details of this example. Although a very simple example, it illustrates how one proceeds in more complicated examples. Notice that knowledge of the sex of the child changes the probability of having an affected child from 1/4 (before the sex was known) either to 1/2 (in the case of a son) or to 0 (in the case of a daughter). Conditional probabilities are used extensively in genetic counseling. We shall now discuss the use of a very general theorem that gives us a mechanism for adjusting the probabilities of events as more information becomes available.

## BAYES' THEOREM

The Englishman Thomas Bayes wrote an essay on probability that he was hesitant to publish because he recognized the flaw in assuming, as he did in his essay, that all possible outcomes are equally likely. The essay was nevertheless published in 1763, after his death, by a friend. What is now called "Bayes' theorem" evolved from this essay and does not contain the original flaw. The theorem gives us a method of calculating new probabilities to take account of new information. Suppose, for example, that a woman knows that her mother carries the gene for hemophilia (because her brother and her maternal grandfather both have the disease). She is pregnant with a male fetus and wants to know the probability that he will be born with hemophilia. There is a $1/2$ probability that she has inherited the hemophilia gene from her mother, and a $1/2$ probability that (given she inherited the gene) she passes it on to her son. Thus, if this is the only information available, the probability that her son will have hemophilia is $1/4$. Now suppose we are told that she already has a son who does not have hemophilia. What is now the probability that her next son will be born with hemophilia? Is it the same, is it greater than $1/4$, or is it less than $1/4$? That she has already had a son without hemophilia is new information that has a direct bearing on the situation. In order to see this, consider a more extreme situation: suppose she has already had 10 unaffected sons. This would suggest that she did not inherit the gene from her mother, and if that is the case she cannot pass it on to her future sons. If, on the other hand, she already has had a son with hemophilia, she would know without a doubt that she had inherited the gene from her mother and every future son would have a $1/2$ probability of being affected. Thus, the fact that she has had one son who is unaffected **decreases** the probability that her second son will be affected. We shall now state Bayes' theorem, which gives us a formula for calculating the new probability.

Let the new information that is available be that the event B has occurred. (In our example, B is the event that there is a previous son without hemophilia.) Now suppose the event B can occur in any one of k distinct, mutually exclusive, ways. Call these ways $C_1$, $C_2$, . . . , $C_k$ (in our example there are two ways of interest: followed by a son with hemophilia or followed by a son without hemophilia), and that with no knowledge about B these have probabilities $P(C_1)$, $P(C_2)$, . . . , and $P(C_k)$, respectively. These probabilities are called the **a priori,** or simply **prior,** probabilities (i.e., prior to our knowledge that event B has occurred). Then the theorem states that the probability of a particular C, say $C_j$, conditional on B having occurred, is

$$P(C_j | B) = \frac{P(C_j)P(B | C_j)}{P(C_1)P(B | C_1) + P(C_2)P(B | C_2) + . . . + P(C_k)P(B | C_k)}.$$

This is called the **a posteriori,** or simply **posterior,** probability of $C_j$

(i.e., the probability **after** it is known that B has occurred). Since $P(C_j)P(B|C_j) = P(C_j \text{ and } B)$, the theorem can also be written

$$P(C_j|B) = \frac{P(C_j \text{ and } B)}{P(C_1 \text{ and } B) + P(C_2 \text{ and } B) + \ldots + P(C_k \text{ and } B)}.$$

It can thus be remembered as "the joint probability divided by the sum of the joint probabilities" (i.e., the posterior probability of a particular C, given that B has occurred, is equal to the joint probability of C and B occurring, divided by the sum of the joint probabilities of each of the C's and B occurring). Intuitively, the theorem is saying that since we know that B has occurred, our new probability for C should be the ratio of the probability of C and B to the probability of any possible C and B.

Now let us apply this theorem to our example. We have $k = 2$ and define the following events:

$$B = \text{first son is unaffected}$$
$$C_1 = \text{second son is affected}$$
$$C_2 = \text{second son is unaffected}$$

If we did not know that the first son was unaffected, we would have the prior probabilities:

$$P(C_1) = \frac{1}{4}, \quad P(C_2) = \frac{3}{4}.$$

(Notice that the prior probabilities must add up to 1; otherwise, we do not have a complete set of true probabilities.) We shall use the second expression for the theorem to calculate the probability that a son is affected after we know that the mother has had one unaffected son. We need to calculate:

$$P(C_1|B) = \frac{P(C_1 \text{ and } B)}{P(C_1 \text{ and } B) + P(C_2 \text{ and } B)}.$$

Now the events $C_1$ and B are not independent, as we have seen. They are, however, independent if we know the woman is a carrier of the hemophilia gene, or alternatively if we know that she is not a carrier. If she is a carrier, the probability of her first son being unaffected and her second son being affected is $1/2 \cdot 1/2 = 1/4$, or,

$$P(C_1 \text{ and } B \mid \text{woman is a carrier}) = \frac{1}{2} \cdot \frac{1}{2} = \frac{1}{4}.$$

If, on the other hand, she is not a carrier, the probability of her first son being unaffected and her second son being affected is $1 \cdot 0 = 0$. (The woman cannot have a second son who is affected if she is not a carrier; we ignore the possibility of a mutation.) Thus

$$P(C_1 \text{ and } B \mid \text{woman is not a carrier}) = 1 \cdot 0 = 0.$$

The woman must either be a carrier (with probability $1/2$) or not a carrier (with probability $1/2$), and so we have

$$
\begin{aligned}
P(C_1 \text{ and } B) &= P(\text{woman is a carrier and } C_1 \text{ and } B) \\
&\quad + P(\text{woman is not a carrier and } C_1 \text{ and } B) \\
&= \frac{1}{2} P(B \text{ and } C_1 \mid \text{woman is a carrier}) \\
&\quad + \frac{1}{2} P(B \text{ and } C_1 \mid \text{woman is not a carrier}) \\
&= \frac{1}{2} \left( \frac{1}{4} \right) + \frac{1}{2} (0) = \frac{1}{8}.
\end{aligned}
$$

This argument is illustrated by a tree diagram in Figure 4.2.
 Similarly, we have

$$
\begin{aligned}
P(C_2 \text{ and } B) &= \frac{1}{2} P(B \text{ and } C_2 \mid \text{woman is a carrier}) \\
&\quad + \frac{1}{2} P(B \text{ and } C_2 \mid \text{woman is not a carrier}) \\
&= \frac{1}{2} \left( \frac{1}{2} \cdot \frac{1}{2} \right) + \frac{1}{2} (1 \cdot 1) = \frac{5}{8}.
\end{aligned}
$$

So we find

$$P(C_1 \mid B) = \frac{\frac{1}{8}}{\frac{1}{8} + \frac{5}{8}} = \frac{1}{6}.$$

Thus, whereas before we know she has already had one unaffected son, the probability of a son being hemophilic is $1/4$, the knowledge of

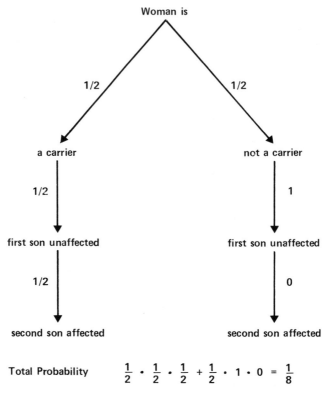

**FIGURE 4.2** Calculation of $P(C_1$ and B), the probability that a woman who has probability 1/2 of being a carrier for hemophilia should have her first son unaffected and her second son affected with hemophilia.

that fact reduces the probability to 1/6. In general, the more unaffected sons she has, the less the probability that her next son will be affected.

Let us take as another example a situation that could arise in paternity testing. Suppose a mother has blood type A (and hence genotype AA or AO), and her child has blood type AB (and hence genotype AB). Thus, the child's A gene came from the mother and the B gene must have come from the father. The mother alleges that a certain man is the father, and his blood is typed. Consider two possible cases:

## CASE 1

The alleged father has blood type O, and hence genotype OO. Barring a mutation, he could not have been the father: this is called an exclusion. When there is an exclusion we do not need any further analysis.

## CASE 2

The alleged father has blood type AB, and therefore genotype AB, which is relatively rare in the population. Thus, not only **could** the man be the father, but he is also **more likely** to be the father than a man picked at random (who is much more likely to be O or A, and hence not be the father). In this case we might wish to determine, on the basis of the evidence available, the probability that the alleged father is in fact the true father. We can use Bayes' theorem for this purpose, provided we are prepared to make certain assumptions:

1. Assume there have been no blood-typing errors and no mutation, and that we have the correct mother. It follows from this assumption that the following two events occurred: (1) the alleged father is AB; and (2) the child received B from the true father—call this event B.
2. Assume we can specify the different, mutually exclusive ways in which B could have occurred. For the purposes of this example we shall suppose there are just two posible ways in which event B could have occurred (i.e., $k = 2$, again in the theorem):

    (1) the alleged father is the true father ($C_1$). We know from Mendelian genetics that the probability of a child receiving a B gene from an AB father is 0.5, i.e., $P(B|C_1) = 0.5$; or

    (2) a random man from a specified population is the true father ($C_2$). We shall assume that the frequency of the gene B in this population is 0.06, so that $P(B|C_2) = 0.06$. The probability that a B gene is passed on to a child by a random man from the population is the same as the probability that a random gene in the population is B (i.e., the population gene frequency of B).
3. Assume we know the prior probabilities of these two possibilities. As explained in Appendix 1, these could be estimated from the previous experience of the laboratory doing the blood typing. We shall assume it is known that 65% of alleged fathers whose cases are typed in this laboratory are in fact the true

fathers of the children in question, i.e., $P(C_1) = 0.65$ and $P(C_2)$
$= 0.35$.

We are now ready to use the formula

$$P(C_1 | B) = \frac{P(C_1)P(B|C_1)}{P(C_1)P(B|C_1) + P(C_2)P(B|C_2)} ,$$

substituting $P(C_1) = 0.65$, $P(C_2) = 0.35$, $P(B|C_1) = 0.5$ and $P(B|C_2)$
$= 0.06$. Thus we obtain

$$P(C_1 | B) = \frac{(0.65)(0.5)}{(0.65)(0.5) + (0.35)(0.06)}$$

$$= \frac{0.325}{0.325 + 0.021}$$

$$= 0.94 .$$

A summary of this application of Bayes' theorem is shown by means
of a tree diagram in Figure 4.3.

FIGURE 4.3 Calculation of the probability that the father of a child is the alleged
father, on the basis of ABO blood types, using Bayes' theorem.

Thus whereas the alleged father, considered as a random case coming to this particular paternity testing laboratory, originally had a 65% probability of being the true father, now, when we take the result of this blood typing into consideration as well, he has a 94% probability of being the true father. In practice, many different blood systems are determined, which usually result in either one or more exclusions or a high probability of paternity. We can never prove that the alleged father is the true father, since a random man could also have the same blood type as the true father. But if enough genetic systems are typed, either an exclusion will be found or the final probability will be very close to unity. In fact it will soon be possible, using the many new genetic systems that are currently being discovered in DNA (obtainable from the white cells of a blood sample), to exclude virtually everyone except the monozygotic twin of the true father.

Several points should be noted about the use of Bayes' theorem for calculating the "probability of paternity." First, when the method was initially proposed, it was assumed that $P(C_1) = P(C_2) = 0.5$ (and in fact, there may be many who still misguidedly make this assumption). The argument given was that the alleged father was either the true father or was not, and in the absence of any knowledge about which was the case, it would seem reasonable to give these two possibilities equal prior probabilities. To see how unreasonable such an assumption is, consider the following. If I roll a die, it will come up either a "six" or "not a six." Since I do not know which will happen, should I assume equal probabilities for each of these two possibilities? In fact, of course, previous experience and our understanding of physical laws suggest that the probability that a fair die will come up a "six" is $1/6$, whereas the probability it will come up "not a six" is $5/6$. Similarly we should use previous experience and our knowledge of genetic theory to come up with a reasonable prior probability that the alleged father is the true father. We can understand now why Thomas Bayes was very unhappy with the idea that lack of knowledge about prior probabilities implies that they are equal.

Second, although the probability of paternity obtained in this way may be perfectly valid, it may not be relevant for the particular man in question. A blood-typing laboratory may validly calculate a 99% probability of paternity for 100 different men, and exactly one of these men may not be the true father of the 100 children concerned.

But if it were known, for that one man, that he could not have had access to the woman in question, the relevant probability for him would be 0. It was a step forward when blood-typing evidence became admissible in courts of law, but it would be a step backward if, on account of this, other evidence were ignored. The so-called probability of paternity summarizes the evidence from the blood-typing only.

Last, always remember that probabilities depend on certain assumptions. At the most basic level, a probability depends on the population of reference. Thus when we assumed $P(B | C_2) = 0.06$, we were implicitly assuming a population of men in which the relative frequency of the B gene is 0.06—which is appropriate for Caucasians but not necessarily for other racial groups. Had we not wished to make this assumption, but rather that the father could have been from one of two different racial groups, it would have been necessary to assume the specific probabilities that he came from each of those racial groups. A posterior probability obtained using Bayes' theorem also assumes we know all the possible ways that the event B can occur, and can specify an appropriate prior probability for each. In our example, we assumed that the true father was either the alleged father—the accused man—or a random Caucasian man. But could the true father have been a relative of the woman? Or a relative of the accused man—perhaps his brother? We can allow for these possibilities when we use Bayes' theorem, since in the general theorem we are not limited to k = 2, but then we must have an appropriate prior probability for each possibility. In practice, it may be difficult to know what the appropriate prior probabilities are, even if we are sure that no relatives are involved.


## LIKELIHOOD RATIO

If we are unwilling to assume particular prior probabilities, it is impossible to derive a posterior probability. But we could measure the strength of the evidence that the alleged father is the true father by the ratio $P(B | C_1)/P(B | C_2) = 0.5/0.06 = 8.3$, which in this particular example is called the "paternity index." It is simply the probability of what we observed (the child receiving allele B from the true father) if the alleged father is in fact the father, **relative to** the probability of what we observed if a random Caucasian man is the father. This is an example of what is known as a **likelihood ratio.** If we have any two hypotheses $C_1$ and $C_2$ that could explain a particular event B, then the likelihood ratio of $C_1$ relative to $C_2$ is defined as the ratio $P(B \mid C_1)/P(B \mid C_2)$. The conditional probability $P(B \mid C_1)$—the probability of observing B if the hypothesis $C_1$ ("the alleged father is the true father") is true—is also called the "likelihood" of the hypothesis $C_1$ on the basis of B having occurred. Similarly, $P(B \mid C_2)$ would be called the likelihood of $C_2$. The likelihood ratio is a ratio of two conditional probabilities, and is used to assess the relative merits of two "conditions" ($C_1$ versus $C_2$), or hypotheses. We shall come back to this concept in Chapter 7.

### SUMMARY

1. The probability of event A is classically defined as the number of outcomes that are favorable to A divided by the total number of possible outcomes. This definition has the disadvantage that it requires one to assume that all possible outcomes are equiprobable. The frequency definition assumes one can perform a trial many times, and defines the probability of A as the limiting proportion, as the number of trials tends to infinity, of the trials in which A occurs. The axiomatic definition of probability is simply a set of positive numbers that sum to unity.

2. A valid probability need not be the clinically relevant probability; the patient at hand may belong to a special subset of the total population to which the probability refers.

3. The probability of A or B occurring is given by

$$P(A \text{ or } B) = P(A) + P(B) - P(A \text{ and } B) .$$

If A and B are mutually exclusive events, $P(A \text{ and } B) = 0$.

4. The joint probability of A and B occurring is given by

$$P(A \text{ and } B) = P(A)P(B \mid A) = P(B)P(A \mid B) .$$

If A and B are independent, $P(A \text{ and } B) = P(A) P(B)$, and conversely, if $P(A$ and $B) = P(A) P(B)$, then A and B are independent. Many different biological mechanisms can be the cause of dependence. Mutually exclusive events are never independent.

5. The conditional probability of A given B is given by

$$P(A|B) = \frac{P(A \text{ and } B)}{P(B)}.$$

6. Bayes' theorem states that the posterior probability that a particular $C_j$ occurred, after it is known that the event B has occurred, is equal to the joint probability of $C_j$ and B divided by the sum of the joint probabilities of each possible C and B:

$$P(C_j|B) = \frac{P(C_j \text{ and } B)}{P(C_1 \text{ and } B) + P(C_2 \text{ and } B) + \ldots + P(C_k \text{ and } B)}$$

$$= \frac{P(C_j)P(B|C_j)}{P(C_1)P(B|C_1) + P(C_2)P(B|C_2) + \ldots + P(C_k)P(B|C_k)}$$

It is assumed that we can specify a complete set of mutually exclusive ways in which B can occur, together with the prior probability of each. It does not assume that these prior probabilities are all equal.

7. When the prior probabilities of $C_1$ and $C_2$ are unknown, we can consider $P(B|C_1)/P(B|C_2)$, the likelihood ratio of $C_1$ versus $C_2$, as a summary of the evidence provided by the event B relative to $C_1$ and $C_2$.

## FURTHER READING

Inglefinger, J.A., Mosteller, F., Thibodeaux, L.A. and Ware, J.H. (1983) Biostatistics in Clinical Medicine. Macmillan. (Chapter 1 gives several good examples of probability applied to clinical cases.)

Weinstein, M.C., Fineberg, H.V., Elstein, A.S., Frazier, H.S., Neuhauser, D., Neutra, R.R. and McNeil, B.J. (1980) Clinical Decision Analysis. W.B. Saunders. (Section 4.9 gives an interesting illustration of Bayes' theorem applied to emergency-ward patients with acute abdominal pain. It is seen how the probability that such a patient has appendicitis, pancreatitis, or nonspecific abdominal disease changes after the result of a simple clinical test, rebound tenderness, is known.)

# PROBLEMS

Problems 1–4 are based on the following:

For a particular population, the lifetime probability of contracting glaucoma is approximately 0.007 and the lifetime probability of contracting diabetes is approximately 0.020. A researcher finds (for the same population) that the probability of contracting both of these diseases in a lifetime is 0.0008.

1. What is the lifetime probability of contracting either glaucoma or diabetes?

2. What is the lifetime probability of contracting glaucoma for a person who has, or will have, diabetes?

3. What is the lifetime probability of contracting diabetes for a person who has, or will have, glaucoma?

Possible answers for problems 1–3 are:

A. 0.0400
B. 0.0278
C. 0.0296
D. 0.0262
E. 0.1143

4. On the basis of the information given, which of the following conclusions is most appropriate for the two events: contracting glaucoma and contracting diabetes?

A. They are independent.
B. They are not independent.
C. They have additive probabilities.
D. They have genetic linkage.
E. They have biological variability.

5. A certain operation has a fatality rate of 30%. If this operation is performed independently on three different patients, what is the probability that all three operations will be fatal?

A. 0.09
B. 0.90
C. 0.009
D. 0.027
E. 0.27

6. If two-thirds of patients survive their first myocardial infarction and one-third of these survivors is still alive 10 years after the first attack, then among all patients who have a myocardial infarction, what proportion will die within 10 years of the first attack?

A. $1/9$

B. $2/9$

C. $1/3$

D. $2/3$

E. $7/9$

7. If 30% of all patients who have a certain disease die during the first year, and 20% of the first-year survivors die before the fifth year, what is the probability an affected person survives past 5 years?

A. 0.50

B. 0.10

C. 0.56

D. 0.06

E. 0.14

8. Suppose that 5 men out of 100 and 25 women out of 10,000 are color-blind. A color-blind person is chosen at random. What is the probability the randomly chosen person is male? (Assume males and females to be in equal numbers.)

A. 0.05

B. 0.25

C. 0.75

D. 0.95

E. 0.99

9. A mother with blood type B has a child with blood type O. She alleges that a man with blood type O is the father of the child. What is the likelihood that the man is the true father, based on this information alone, relative to a man chosen at random from a population in which the frequency of the O gene is 0.67?

A. 0.33

B. 1.49

C. 2.00

D. 0.50

E. 0.67

10. There is a 5% chance that the mother of a child with Down's syndrome has a particular chromosomal translocation, and if she has that translocation, there is a 16% chance that a subsequent child will have Down's syndrome; otherwise the chance of a subsequent child having Down's Syndrome is only 1%. Given these facts, which of the following statements is true for a woman with a Down's syndrome child?

(1) The probability that a subsequent child will have Down's syndrome is independent of whether or not the mother has the translocation.

(2) The joint probability that the mother has the chromosomal translocation and that her next child has Down's syndrome is about 21%.

(3) The probability that her next child will have Down's syndrome is at least 5%.

(4) The probability that her next child will have Down's syndrome is almost 2%.

   A. Only 1, 2, and 3.
   B. Only 1 and 3.
   C. Only 2 and 4.
   D. Only 4.
   E. 1, 2, 3, and 4.

## CHAPTER 5

# Random Variables and Distributions

**Key Concepts:** random variable; probability (density) function, cumulative probability distribution function; binomial, Poisson, uniform, and normal distributions.

## VARIABILITY AND RANDOM VARIABLES

A general term for any characteristic or trait we might measure on a patient or other unit of study is **variable.** A patient's height, weight, and blood pressure are examples of variables. These traits are variable (and hence called variables) for two main reasons:

1. Measurement error causes the values we measure to differ, even though we may be repeatedly measuring exactly the same thing. Broadly speaking, measurement error includes errors of misclassification, variability in the measuring instruments we use, and variability among observers in how they read those instruments. For example, no two mercury sphygmomanometers are exactly the same. Even if a group of doctors all view the same film of the movement of mercury in such an instrument, together with listening to the sounds heard in the stethoscope, they will nevertheless not all agree on how they read the diastolic and systolic blood pressures.
2. There is inherent variability in all biological systems. There are differences among species, among individuals within a species, and among parts of an individual within an individual. No two human beings have exactly the same height and weight at all stages of their growth; no two muscle cells have exactly the same chemical composition. The same individual, or the same

muscle cell, changes with age, time of day, season of the year, and so forth. An individual's blood pressure, for example, can fluctuate widely from moment to moment depending on how fearful the subject is.

When a variable is observed as part of an experiment or a well-defined sampling process, it is usually called a **random variable.** Thus, we think of each observation in a set of data as the outcome of a random variable. For example, consider a family in which there is a given probability that a child is affected with some disease. Then whether or not a child is affected is the outcome of a discrete random variable. We often assign arbitrary numbers to the outcomes of such a random variable, for instance, 0 if not affected, 1 if affected. In other settings, variables such as blood pressure measurements and cholesterol levels are continuous random variables.

Data of all types, but especially those that involve observations on a large number of subjects, are difficult to interpret unless they are organized in a way that lends itself to interpretation. A first and important way of organizing data to obtain an overview of the patterns they display is to construct a frequency distribution, as we described in Chapter 3. Here we shall be concerned with **probability distributions.** A probability distribution is a model for a random variable, describing the way the probability is distributed among the possible values the random variable can take on. As we saw in Chapter 4, probability can be interpreted as relative frequency in an indefinitely large number of trials. The distributions we shall now describe are thus theoretical ones, but nevertheless ones that are of great practical importance and utility. Mathematically, the concepts "probability distribution" and "random variable" are interrelated, in that each implies the existence of the other; a random variable must have a probability distribution and a probability distribution must be associated with a random variable.

## BINOMIAL DISTRIBUTION

Let us start with a simple dichotomous trait (a qualitative trait that can take on one of only two values). Consider a rare autosomal dominant condition such as achondroplasia (a type of dwarfism). Let the random variable we are interested in be the number of children with achondroplasia in a family. Suppose a couple, one of whom has achondroplasia, has a single child. Then the random variable is dichotomous (it must be 0 or 1), and the probability the child is unaffected is $1/2$. We thus have

$$P(0) = \frac{1}{2},$$

$$P(1) = \frac{1}{2},$$

which we can graph as in Figure 5.1(a). If we call the random variable Y, we can also write this as

$$P(Y = 0) = \frac{1}{2},$$

$$P(Y = 1) = \frac{1}{2}.$$

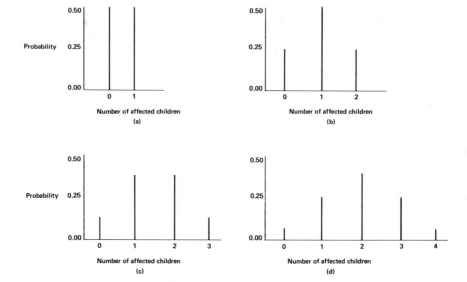

FIGURE 5.1 Probability distribution of number of children affected when the probability of each child being affected is $1/2$ and the total number of children in a family is (a) 1, (b) 2, (c) 3, and (d) 4.

This completely describes the probability distribution of the random variable Y for this situation. Note that the distribution is described by a function of Y. This function, which gives us a rule for calculating the probability of any particular value of Y, is called a **probability function.**

Now suppose the couple has two children; then Y, the number affected, can take on three different values: 0, 1, or 2. (Note this is a discrete random variable measured on an ordinal scale.) Using the laws of probability and the fact that each conception is an independent event (we shall exclude the possibility of monozygotic twins), we have

$$P(0) = P(\text{1st child unaffected and 2nd child unaffected})$$
$$= P(\text{1st child unaffected})P(\text{2nd child unaffected} \mid \text{1st child unaffected})$$
$$= \frac{1}{2} \cdot \frac{1}{2}$$
$$= \frac{1}{4}.$$

$$P(1) = P(\text{one child affected})$$
$$= P(\text{1st child affected and 2nd child unaffected})$$
$$\quad + P(\text{1st child unaffected and 2nd child affected})$$
$$= P(\text{1st child affected})P(\text{2nd child unaffected} \mid \text{1st child affected})$$
$$\quad + P(\text{1st child unaffected})P(\text{2nd child affected} \mid \text{1st child unaffected})$$
$$= \frac{1}{2} \cdot \frac{1}{2} + \frac{1}{2} \cdot \frac{1}{2}$$
$$= \frac{1}{2}.$$

$$P(2) = P(\text{Both children affected})$$
$$= P(\text{1st child affected})P(\text{2nd child affected} \mid \text{1st child affected})$$
$$= \frac{1}{2} \cdot \frac{1}{2}$$
$$= \frac{1}{4}.$$

Note carefully that there are two mutually exclusive ways in which Y can take on the value 1: either the first child is affected **or** the second child is affected, and the other is not. This is why we simply added the two probabilities.

In summary, for a family of two children, we have the probability function

$$P(0) = \frac{1}{4}, \; P(1) = \frac{1}{2}, \; P(2) = \frac{1}{4},$$

which is graphed in Figure 5.1(b).

Using the same kind of argument, we find that the probability function of Y in a family with three children is

$$P(0) = \frac{1}{8}, \; P(1) = \frac{3}{8}, \; P(2) = \frac{3}{8}, \; P(3) = \frac{1}{8};$$

and for a family of four children we find

$$P(0) = \frac{1}{16}, \; P(1) = \frac{1}{4}, \; P(2) = \frac{3}{8}, \; P(3) = \frac{1}{4}, \; P(4) = \frac{1}{16}.$$

These are graphed in Figures 5.1(c) and (d).

All the probability functions we have so far considered are special cases of the **binomial probability distribution,** or simply the **binomial distribution.** The expression "binomial distribution" is used to mean a family of many different distributions, the particular one intended being determined by the values of two (non-random) variables called **parameters.** Suppose we perform n independent trials, and at each trial the probability of a particular event is $\pi$. Then n and $\pi$ are the parameters of the distribution. In the above examples n is the number of children in the family and $\pi$ is the probability that each should have achondroplasia. The probability that the event occurs a total of y times is given by the binomial probability distribution with parameters n and $\pi$, and is expressed mathematically by the probability function

$$P(y) = \frac{n!}{y!(n-y)!} \; \pi^y (1 - \pi)^{n-y}, \text{ for } y = 0, 1, 2, \ldots, n.$$

In this formula, read n! as "n factorial," or "factorial n"; it is equal to $n(n - 1)(n - 2) \ldots (2)(1)$. For example, $3! = (3)(2)(1) = 6$. To use the formula, you may need to use the fact that 0! is defined to be unity (i.e., 0! = 1). Also, remember that any number raised to the power zero is unity [e.g., $(1/2)^0 = 1$]. Consider, for example, finding the probability that, in a family with four children in which one parent is achondroplastic, no children will be affected. Here the parameters are n = 4 and $\pi = 1/2$, and we want to evaluate the probability function for y = 0. Thus

$$P(0) = \frac{4!}{0!(4-0)!} \left(\frac{1}{2}\right)^0 \left(1 - \frac{1}{2}\right)^{4-0}$$
$$= \frac{4!}{4!}(1) \left(\frac{1}{2}\right)^4$$
$$= \left(\frac{1}{2}\right)^4 = \frac{1}{16}.$$

Note that in general, $P(0) = (1 - \pi)^n$, that is, the probability of no events of the type being considered occurring in a total of n trials is $(1 - \pi)^n$. Conversely, the probability of more than zero (i.e., at least one) such events occurring must be the complement of ("complement of" means "one minus") this, that is, $1 - (1 - \pi)^n$: the number of such events occurring must be either zero or more than zero. Finally, remember that the binomial probability distribution is relevant for a prespecified number n of **independent** trials in each of which there is the **same** probability $\pi$ of a particular event occurring. Binomial probabilities are often relevant in genetic counseling because each conception can be considered an independent event with the same probability of resulting in a child having a given condition.

## A NOTE ABOUT SYMBOLS

You may have noticed that we originally used the capital letter Y to denote a random variable, but when we gave the formula for the binomial distribution we used the small letter y. Throughout this book we shall use capital Roman letters for random variables and the corresponding small Roman letters to denote a particular value that the random variable takes on. Thus P(y) is a shorthand way of writing P(Y = y), the probability that the random variable Y takes on the particular value y (such as 0, 1, 2, etc.). Making this distinction now will make it easier to understand some of the concepts we discuss in later chapters. Just remember that a capital letter stands for a random variable, which can take on different values, whereas a lowercase letter stands for a particular number. We shall also, for the most part, use Greek letters for unknown population parameters. Thus we shall use p for the proportion of affected children we might observe in a sample of children, but $\pi$ for the theoretical probability that a child should be affected. (Since it is not universal, however, this convention will not be followed in all the problems.) The Greek letter $\pi$ will also be used later in this chapter with its usual mathematical meaning of the ratio of the circumference of a circle to its diameter. In the next section we shall need to use the theoretical quantity that mathematicians always denote by the Roman letter e. Like $\pi$, e cannot be expressed as a rational number; it is approximately 2.71828.

## POISSON DISTRIBUTION

Another important family of discrete distributions is the **Poisson distribution,** named after the French mathematician S.D. Poisson (1791–1840). This can often be used as a model for random variables such as radioactive counts per unit of time, the number of calls arriving at a telephone switchboard per unit of time, or the number of bacterial colonies per Petri plate in a microbiology study. If y is a particular value of a random variable that follows a Poisson distribution, then the probability that this value will occur is given by the formula

$$P(y) = \frac{\lambda^y e^{-\lambda}}{y!} \text{ for } y = 0, 1, 2, \ldots$$

where the parameter $\lambda$ is the mean or average value of the random variable.

Whereas the binomial is a two-parameter (n and $\pi$) family of distributions, the Poisson is a one-parameter ($\lambda$) family. The Poisson distribution can be derived as a limiting case of the binomial, and is sometimes called the distribution of "rare events." Suppose we have a binomial distribution but the number of trials, n, is indefinitely large while the probability $\pi$ of a particular event at each trial approaches zero; the resulting distribution, provided the average number of events expected in n trials is a finite quantity, is the Poisson. Thus the Poisson distribution also assumes independent events, each with the same probability of occurring, but in addition it assumes that the total number of such events **could be** (though with very small probability) indefinitely large.

As an example, suppose it is known that in a large hospital, two patients a month, on an average, give incorrect billing information. What is the probability of a month with no patient giving incorrect billing information in this hospital? Or that there should be one patient with incorrect billing information in a given month? Or two, or three? In this example $\lambda = 2$. Thus, using the above formula, we can calculate

$$P(0) = \frac{2^0 e^{-2}}{0!} = 0.1353$$

$$P(1) = \frac{2^1 e^{-2}}{1!} = 0.2706$$

$$P(2) = \frac{2^2 e^{-2}}{2!} = 0.2706$$

$$P(3) = \frac{2^3 e^{-2}}{3!} = 0.1804$$

and so on. Note that the probability of no such events is in general $e^{-\lambda}$, and the probability of at least one such event is $1 - e^{-\lambda}$.

Poisson probabilities are good approximations to binomial probabilities when n is large. Suppose, for example, that in the same hospital exactly 500 patients are billed each month. Then it would be appropriate to use the binomial distribution with n = 500 and $\pi$ = 2/500, rather than the Poisson distribution. We would then find, for no incorrect billings,

$$P(0) = \left( 1 - \frac{1}{250} \right)^{500}$$

$$= \left( \frac{249}{250} \right)^{500} = 0.1348 \,,$$

so we see that the Poisson approximation (0.1353) is good to three decimal places in this instance. The Poisson distribution is appropri-

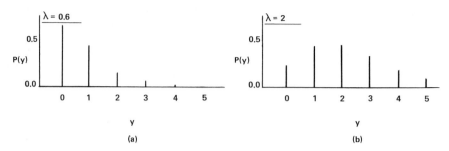

FIGURE 5.2 Examples of the Poisson distribution with parameter λ.

ate when n is truly indefinitely large and so we have no idea what it is. This would be the case when we are measuring the number of radioactive counts in a unit of time, or when we are counting the number of red cells that fall in a square on a hemocytometer grid. Two examples of the Poisson distribution are shown in Figure 5.2. The first ($\lambda = 0.6$) gives a good approximation to the distribution of the number of soldiers in the Prussian army killed by horsekicks in any one year!

## UNIFORM DISTRIBUTION

The two families of distributions we have considered so far—the binomial and the Poisson—are for discrete random variables. In each case we had a formula for $P(y)$, the probability that the random variable Y takes on the value y, and we could plot these probabilities against y. A difficulty arises when we try to do this for a continuous random variable.

Consider as a model a spinner connected to a wheel that has equally spaced numbers ranging from 1 to 12 on its circumference, as on a clock (see Figure 5.3): If you spin the spinner (arrow), its tip is equally likely to stop at any point in the interval from 1 to 12 or, for that matter, at any point between any two consecutive numbers, say between 1 and 2. There are so many points on this circle that the probability the tip stops at any specific point is virtually zero. Yet, it is easy to see that the probability the tip stops between 1 and 2 is the same as the probability it stops between 9 and 10 (and, in fact, this probability is 1/12). We can plot the probability distribution corre-

**FIGURE 5.3** Spinner wheel for a uniform distribution.

sponding to this model as in Figure 5.4, in which the total area of the rectangle is unity: area = base × length = 12 (¹/₁₂) = 1. Moreover, the portion of the total area that lies between 1 and 2 is ¹/₁₂, and this is the probability the tip of the spinner stops between 1 and 2. Similarly, the area between 3 and 7 is 4(¹/₁₂) = ¹/₃, and this is the probability the spinner stops betwen these two numbers. The height of the continuous line (¹/₁₂) is called the **probability density.**

The distribution described by the above model is called the **uniform** (or rectangular) distribution. It is one of the simplest for continuous data. We write the formula for the probability density as follows:

$$f(y) = {}^{1}/_{12} , 0 \le y \le 12 .$$

The random variable Y is the exact point between 0 and 12 at which the tip of the arrow comes to rest. The **probability density function** f(y) gives the height of the line plotted at each value y that Y can take on (in this case the height is always the same, ¹/₁₂, for all values of y between 0 and 12). We cannot call the height the probability of y

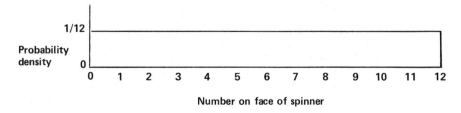

**FIGURE 5.4** The uniform probability density corresponding to the spinner model in Figure 5.3.

because, as we have noted, the probability of any particular value of y—such as 2.5 or 7.68—is essentially zero. It is nevertheless a probability function, but we call it a probability density function, or simply a **density,** to stress that its values are not actual probabilities. The important thing to remember is that in the case of a continuous random variable it is the **areas** under the probability density function f(y) that represent probabilities.

## NORMAL DISTRIBUTION

The most important family of continuous densities used in statistical applications is the **normal** or **Gaussian.** The latter name comes from the German astronomer K. F. Gauss (1777–1855). The density has a bell-shaped appearance, and can also be shown to have a shape identical to that of a binomial distribution as n, the number of trials, becomes indefinitely large while $\pi$ remains constant. Look again at the binomial distributions for $\pi = 1/2$ and n = 1, 2, 3, and 4 in Figure 5.1, and then at the one in Figure 5.5(a) for $\pi = 1/2$ and n = 16. You will see that the shape is approaching the curve in Figure 5.5(b), which is a normal density.

The normal probability density function is symmetric (and hence has a coefficient of skewness equal to zero), with the highest point at the center. In Figure 5.5(b) the highest point occurs when y = 8, and this is the mean of the distribution—denoted by $\mu$, the Greek letter

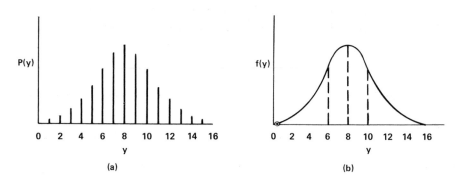

**FIGURE 5.5** (a) The binomial probability function when $\pi = 1/2$ and n = 16. (b) The normal density function when $\mu = 8$ and $\sigma = 2$.

mu. Although the normal density is always bell-shaped, it doesn't always have exactly the same amount of spread: it may be tall and thin, or it may be short and fat. It always, however, has two points of inflection, symmetrically placed about the mean. In Figure 5.5(b) these occur at y = 6 and y = 10; they are points at which the curve changes from being convex to concave, or from concave to convex. Starting at the extreme left, the curve rises more and more steeply until it reaches y = 6, at which point, although still rising, it does so more and more slowly. Similarly, at y = 10 the curve changes from decreasing more and more steeply to decreasing less and less steeply. The distance between the mean and the point of inflection (2 in Figure 5.5) measures the amount of spread of a normal density and is denoted by σ, the Greek letter sigma; it is the standard deviation of the distribution. The family of normal densities is thus a two-parameter family. The general formula for the curve is

$$f(y) = \frac{e^{-\frac{1}{2}\left(\frac{y-\mu}{\sigma}\right)^2}}{\sigma\sqrt{(2\pi)}} \text{ , for } -\infty \leq y \leq \infty,$$

where π and e are the usual mathematical constants, y the value of the random variable, and μ and σ the two parameters. Recall that ∞ is the mathematical symbol for "infinity"; thus the curve goes on "forever" in both directions. The properties that we have discussed, as well as those we shall discuss below, can all be deduced mathematically from this formula.

As with any other density, the total area under the normal curve is unity. Because it is symmetric, the probability to the left of the mean is 1/2, as is the probability to the right of the mean. The mean is thus the same as the median. The normal distribution has two other properties that you should commit to memory:

1. The probability that a normally distributed random variable lies within one standard deviation of the mean, that is, between μ − σ and μ + σ [between 6 and 10 in Figure 5.5(b)], is about two-thirds, or to be a little more precise, about 68%. It follows that there is a 16% probability it will lie below μ − σ and a 16% probability it will lie above μ + σ (16 + 68 + 16 = 100).
2. The probability that a normally distributed random variable

lies within two standard deviations of the mean, that is, between $\mu - 2\sigma$ and $\mu + 2\sigma$ [between 4 and 12 in Figure 5.5(b)], is about 95%. It follows that there is about a $2^{1}/_{2}$% probability it will lie below $\mu - 2\sigma$, and the same probability it will lie above $\mu + 2\sigma$ ($2^{1}/_{2} + 95 + 2^{1}/_{2} = 100$).

Theoretically, a normally distributed random variable can take on any value, positive or negative. Thus for the normal density pictured in Figure 5.5(b), for which $\mu = 8$ and $\sigma = 2$, it is theoretically possible (though highly improbable) that Y could be less than $-1,000$ or greater than 2,000. The actual probability however, of being more than four or five standard deviations from the mean is virtually nil, and for all practical purposes can be ignored. For this reason we can often use the normal distribution to approximate real data, such as the heights of men, that we know cannot possibly be negative. Of course, there is nothing abnormal about a random variable not being normally distributed—remember the word "normal" in this context means only that the density follows the bell-shaped curve that is mathematically defined above. But the normal distribution is of great practical importance in statistics because many types of data **are** approximately normally distributed, or can be transformed to another scale (such as by taking logarithms or square roots) on which they are approximately normally distributed. Furthermore, even if the random variable of interest is not normally distributed and a normalizing transformation cannot readily be found, sample averages of the random variable tend to become normally distributed as the sample size increases. This important result holds true even if the random variable is of the categorical type. This result plays a fundamental role in many statistical analyses, since we often reduce our data to averages in an effort to summarize our findings. We shall come back to this important point later.

## CUMULATIVE DISTRIBUTION FUNCTIONS

So far, we have described various distributions by probability functions we have denoted P(y) or f(y)—the former for discrete random variables (e.g., binomial, Poisson) and the latter for continuous random variables (e.g., uniform, normal). These same distributions can also be described by a different kind of function, namely the **cumulative probability distribution function,** or simply the **cumulative distribution.** Statisticians describe distributions both ways, and it is important to know which is being used. In this book we shall always include the word "cumulative," writing "cumulative probability distribution function" or simply "cumulative distribution" when that is meant. It will be understood that if the word "cumulative" is not stated, then a probability function is meant. (This is the usual convention in the articles you are likely to read, but not the convention in mathematical articles).

For every probability function there is exactly one corresponding cumulative distribution function, and vice versa. Thus each gives us exactly the same information about the distribution of a random variable—they are two ways of expressing the same information. They are different mathematical functions, however, and they look quite different when we plot them. Whereas a probability function gives the probability or density of a particular value y of Y (it is analogous to a histogram or frequency polygon), the cumulative distribution function gives the probability Y is less than or equal to a particular value y (it is analogous to a cumulative plot). The cumulative distribution function is particularly useful for continuous random variables, for which we cannot talk of the "probability of y."

The cumulative distribution functions for the uniform distribution illustrated in Figure 5.4 and the normal distribution illustrated in Figure 5.5(b) are depicted in Figure 5.6; they are denoted F(y).

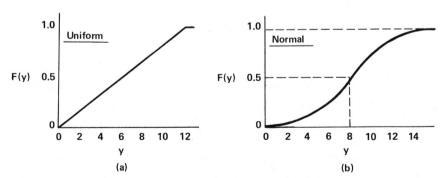

**FIGURE 5.6** Cumulative distribution functions: (a) corresponding to the uniform density function in Figure 5.4; (b) corresponding to the normal density function in Figure 5.5.

First look at the cumulative uniform distribution. The probability that Y is less than or equal to 0 is 0, and we see $F(0) = 0$. The probability that Y lies between 0 and 1 (i.e., is less than or equal to 1) is $1/12$, and we see $F(1) = 1/12$. The probability Y is less than or equal to 6 is $1/2$, and we see $F(6) = 1/2$. Finally, Y must be less than or equal to 12, and in fact we see $F(12) = 1$. Although it is not plotted in the figure, clearly $F(-1) = 0$ and $F(13) = 1$. For any cumulative distribution it is always true that $F(-\infty) = 0$ and $F(\infty) = 1$.

Now look at the cumulative normal distribution illustrated in Figure 5.6(b). Note that, approximately,

$$\begin{aligned} F(4) &= 0.025 \\ F(6) &= 0.16 \\ F(8) &= 0.5 \\ F(10) &= 0.84 \\ \text{and } F(12) &= 0.975 \ . \end{aligned}$$

This illustrates the important properties of the normal distribution you have already learned (look back at the two important properties discussed for the normal distribution and make sure you understand how what was written there in words corresponds to the following equations)

1. $F(\mu + \sigma) - F(\mu - \sigma) = 0.84 - 0.16 = 0.68$ .
2. $F(\mu + 2\sigma) - F(\mu - 2\sigma) = 0.975 - 0.025 = 0.95$ .

We shall be using the cumulative normal distribution, and other cumulative distributions, in later chapters.

### SUMMARY

1. Traits are called variables for two main reasons: (1) Measurement error causes the values we observe to vary when measured repeatedly under the same conditions; and (2) all biological systems are dynamic and hence vary inherently. A random variable is just a variable observed as part of an experiment, or a well-defined sampling process, so that it can be associated with a probability distribution.

2. The distribution of a random variable is a description of the way the probability is distributed among the possible values the ran-

dom variable can take on. There are two ways a distribution can be described: by a probability function (a probability density function in the case of a continuous random variable), or by a cumulative probability distribution function. These each provide the same information about a random variable, but in a different way. A probability function is a rule that gives the probabilities (or densities) associated with different values of the random variable. A cumulative distribution function gives cumulative probabilities.

3. The binomial distribution is relevant for situations in which there are n observations of a dichotomous random variable, the n values arising independently of one another and each observation having the same probability of falling into a given category of the dichotomy; the total number of observations of one type is then binomially distributed.

4. The Poisson distribution can often be used as a model for random variables such as counts of rare events. It is a limiting case of the binomial distribution as the number of observations n becomes indefinitely large.

5. The uniform distribution (density) has a rectangular shape and is characterized by the fact that within well-defined limits, the probability of observing a value of the random variable in any interval is the same as that for any other interval of the same length.

6. The normal distribution (density) is bell-shaped and has a single peak. It is symmetric about its mean $\mu$. The probability that a normally distributed random variable lies within one standard deviation of the mean is 68%; within two standard deviations, 95%.

7. A cumulative distribution gives the probability that a random variable is less than or equal to a particular value. The cumulative normal distribution has the shape of a sloping, elongated "S."

## FURTHER READING

Tsokos, C.P. (1972). Probability Distributions: An Introduction to Probability Theory with Applications. Duxbury Press. (This book provides a mathematical description of many of the distributions likely to be encountered in a variety of applied problems. It is a good starting place for the reader who would like to know more about the mathematical aspects of these distributions.)

## PROBLEMS

1. Which of the following characteristics belong to the normal distribution?
(1) Symmetric.
(2) Mean equal to median.
(3) Median equal to mode.
(4) Positively skewed.
   A. Only 1, 2, and 3.
   B. Only 1 and 2.
   C. Only 2 and 4.
   D. Only 4.
   E. 1, 2, 3, and 4.

2. A physician sent a sample of a patient's blood to a lipid laboratory for cholesterol and triglyceride determinations. The triglyceride determination was well within the normal range, but the cholesterol reading was 280 mg/100 ml. Being unconvinced that the patient had elevated cholesterol, the physician sent a second sample to the laboratory and the reading was 220. The disagreement between the two readings could be an example of
   A. the Poisson distribution
   B. observer error
   C. biological variability
   D. A and B
   E. B and C

3. If we conduct n trials the outcome of each of which is either a success (with probability p) or failure (with probability $1-p$), the distribution of r, the total number of successes, is written

$$P(r) = \frac{n!}{(n-r)!r!} \, p^r(1-p)^{n-r} .$$

This is known as
   A. the binomial distribution with parameters r and p
   B. the binomial distribution with parameters n and p
   C. the binomial distribution with parameters n and r
   D. the binomial distribution with probability p
   E. the binomial distribution with mean np

4. If there is a 2% chance that a child will be born with a congenital anomaly, what is the probability that no congenital anomaly will be found among four random births?

A. 0.92
B. 0.8
C. 0.2
D. 0.02
E. 0.08

5. A couple is at risk of having children with a recessive disease, there being a probability of 0.25 that each child is affected. What is the probability, if they have three children, that at least one will be affected?

A. 0.75
B. 0.02
C. 0.58
D. 0.42
E. 0.25

6. To determine whether mutations occur independently and with equal probability, a researcher sets up an experiment in which the number of mutant bacteria that appear in a certain volume of cell suspension can be counted as the number of colonies on an agar plate. One hundred cell suspensions are plated, and on each of the 100 plates the number of mutant colonies ranges from 0 to 9. In view of the purpose of the study, the distribution of these numbers should be compared to a

A. binomial distribution
B. Poisson distribution
C. uniform distribution
D. normal distribution
E. none of the above

7. The normal distribution has two points of inflection. If the total area is 100%, what is the area under the normal curve between the two points of inflection?

A. 99%
B. 95%

C. 90%
D. 68%
E. 50%

8. A normal distribution has mean 15 and standard deviation 3. What interval includes about 95% of the probability?
   A. 12–18
   B.  9–21
   C.  6–24
   D.  3–27
   E. None of the above.

9. Weight is approximately normally distributed with mean 150 lb. and standard deviation 10 lb. Which of the following intervals includes approximately two-thirds of all weights?
   A. 145–155 lb.
   B. 140–160 lb.
   C. 130–170 lb.
   D. 130–150 lb.
   E. 150–170 lb.

10. A cumulative distribution
   A. expresses the same information as a probability or density function, but in a different way
   B. states the probability that a random variable is less than or equal to a particular value
   C. always takes on a value between zero and one
   D. all of the above
   E. none of the above

CHAPTER 6

# Estimates and Confidence Limits

**Key Concepts:** estimate, estimator; standard error (of the mean); (minimum variance) unbiased, efficient, robust; maximum likelihood; normal range, confidence limits or interval; standardized random variable, t distribution, degrees of freedom; pooled variance.

## ESTIMATES AND ESTIMATORS

It will be helpful at this point to distinguish between an estimate and an estimator. An **estimator** is a rule that tells us how to determine from any sample a numerical value to estimate a certain population parameter, whereas an **estimate** is the actual numerical value obtained from a particular sample. Suppose we select a random sample of 10 students from a class of 200, measure their heights and find the sample mean. If we consider the "sample mean" to be the rule that tells us to add up the 10 heights and divide by 10, it is an estimator, and we shall denote it $\overline{Y}$. If, on the other hand, we consider the "sample mean" to be the result we obtain from a particular sample—say 70 inches—it is an estimate and we shall denote it $\overline{y}$. The estimator $\overline{Y}$ is a random variable that takes on different values from sample to sample. The estimate $\overline{y}$, on the other hand, is the numerical value obtained from one particular sample. We say the sample mean $\overline{Y}$ is an estimator, and $\overline{y}$ an estimate, of the population mean $\mu$. Thus, particular values of an estimator are estimates. Different samples might yield $\overline{y} = 69$ inches or $\overline{y} = 72$ inches as estimates of the mean of the entire class of 200 students, but the estimator $\overline{Y}$ is always the same: add up the 10 heights and divide by 10.

Note we have retained our convention of using capital letters to denote random variables (estimators) and lower-case letters to denote specific values of random variables (estimates). The word "estimate" is often used in both senses, and there is no harm in this provided you understand the difference. To help you appreciate the difference, however, we shall be careful to use the two different words and symbols, as appropriate.

Since an estimator is a random variable, and hence has a distribution, it is of interest to determine its variance and standard deviation. By selecting many different samples and obtaining the corresponding estimate for each sample, we obtain a set of data (estimates) from which we can compute a variance and standard deviation in the usual way. The problem with this approach is that it requires that many samples be studied, and is therefore not practical. Fortunately, however, there is a way of estimating the variance and standard deviation of an estimator from the information available in a single sample. If the population variance of a random variable is $\sigma^2$, it can be proved mathematically that the sample mean $\overline{Y}$ of n independent observations has variance equal to $\sigma^2/n$ and standard deviation equal to $\sigma/\sqrt{n}$. We can estimate these quantities by substituting s for $\sigma$, where s is computed from a single sample in the usual way. We see immediately from the formula $\sigma^2/n$ that the larger the sample size (n) the smaller is the variance and standard deviation of the sample mean. In other words, the sample mean calculated from large samples varies less, from sample to sample, than the sample mean calculated from small samples.

It is usual practice to call the standard deviation of an estimator the **standard error** (often abbreviated s.e.) of that estimator. The standard deviation of the sample mean ($\sigma/\sqrt{n}$) is, therefore, often called the **standard error of the mean.** When applied to the sample mean, there is no real difference between the terms "standard deviation" and "standard error," but the latter is often used when we are referring to variability due to error, as opposed to natural variability. The fact that men have variable heights is in no way due to "error," but the fact that our estimate of mean height differs from the true mean can be considered an error. Furthermore, from now on in this book, as in so much of the medical literature, we shall use the term standard error to indicate the **estimated** standard deviation of the estimator. Thus, the standard error of the mean (often abbreviated s.e.m.) is $s/\sqrt{n}$.

## NOTATION FOR POPULATION PARAMETERS, SAMPLE ESTIMATES AND SAMPLE ESTIMATORS

We have already indicated that we denote population parameters by Greek letters and sample statistics by Roman letters. The former are always fixed constants, whereas the latter, in the context of repeated sampling, can be considered random variables. The particular estimates calculated from one sample (such as 69 inches or 72 inches in the above example) are fixed constant quantities, but they can be viewed as particular outcomes of random variables in the context of examining many, many samples. Depending on how they are viewed, we use lower-case or upper-case Roman letters. Thus, assuming a sample of size n, our notation is as follows:

| Name | Parameter | Estimate | Estimator |
|------|-----------|----------|-----------|
| Mean | $\mu$ | $\bar{y}$ | $\bar{Y}$ |
| Variance | $\sigma^2, \sigma_Y^2$ | $s^2, s_Y^2$ | $S^2, S_Y^2$ |
| Standard deviation | $\sigma, \sigma_Y$ | $s, s_Y$ | $S, S_Y$ |
| Standard deviation of the mean | $\sigma_{\bar{Y}} = \sigma_Y/\sqrt{n}$ | $s_{\bar{y}} = s_Y/\sqrt{n}$ | $S_{\bar{Y}} = S_Y/\sqrt{n}$ |

Thus $\mu$ is the population mean (parameter), $\bar{y}$ the sample mean (estimate). Similarly, $\sigma$ is the population standard deviation, s the sample standard deviation; in other words, $\sigma$ is the standard deviation of a random variable $\bar{Y}$, and s is the estimate of $\sigma$ we calculate from a sample. When we want to stress the fact that the relevant random variable is $\bar{Y}$, we write $\sigma_Y$ or $s_Y$—the standard deviation of $\bar{Y}$. Analogously, $\sigma_{\bar{Y}}$ is the standard deviation of $\bar{Y}$, the standard deviation of the sample mean of n observations, while $s_{\bar{y}}$ is the sample estimate of this quantity, which, as noted earlier, is often called the **standard error of the mean**. Thus $\sigma_{\bar{Y}} = \sigma_Y/\sqrt{n}$ and $s_{\bar{y}} = s_Y/\sqrt{n}$.

## PROPERTIES OF ESTIMATORS

We usually **estimate** the population mean by the sample mean; and we often **estimate** the population variance by the sample variance.

Why do we do this? Why not, for example, estimate the population mean by the average of the smallest and largest values in our sample? The answer is that we choose estimators, and hence estimates, that have certain "good" properties. An estimator is a random variable and has a distribution. The characteristics of this distribution determine the goodness of the estimator.

In the previous example, suppose we take all possible samples of 10 students from the class of 200 and compute the mean height from each sample. These sample means vary, and if we average them we obtain a "mean of means" which is equal to the population mean— the mean height from all 200 students. In a case such as this, in which the average of the sample estimates for all possible samples equals the value of the parameter being estimated, we say the estimator is **unbiased:** the mean of the estimator's distribution is equal to the parameter being estimated. When this does not occur, the estimator is **biased.** In general, when we have a random sample, the sample mean is an unbiased estimator of the population mean and the sample variance is an unbiased estimator of the population variance; but the sample standard deviation is not an unbiased estimator of the population standard deviation.

Recall that we use $n-1$ rather than n as a divisor when we average the sum of squared deviations from the sample mean in computing the variance. If we used n as a divisor, we would find that the average of all possible estimates is equal to $(n-1)/n$ times the population variance and is thus biased; such an estimator leads to an estimate that is, on an average, smaller than the population value. With $n-1$ as a divisor, we have an unbiased estimator of the variance.

We should also like our estimate to be close to the parameter being estimated as often as possible. It is not very helpful for an estimate to be correct "on an average" if it fluctuates widely from sample to sample. Thus, for an estimator to be good its distribution should be concentrated fairly closely about the true value of the parameter of interest; in other words, the variance of the estimator's distribution should be small. If we have a choice among several estimators that are competing, so to speak, for the job of estimating a parameter, we might proceed by eliminating any that are biased and then, from among those that are unbiased, choose the one with the smallest variance. Such an estimator is called a **minimum variance unbiased estimator,** or an **efficient estimator.** It can be shown mathematically that if the underlying population is normally distributed, the sample mean and the sample variance are minimum variance unbiased estimators of the population mean and the population variance, respectively. In other situations, however, the sample descriptive statistic may not have this property. The sample mean is a minimum variance unbiased estimator of the population mean when the underlying population is binomially or Poisson-distributed, but in these two cases an analogous statement cannot be made about the sample variance. In fact, the efficient estimator of the variance of a Poisson distribution is provided by the sample mean!

We have discussed only two important properties of good estimators. Other properties are also of interest, but the important thing is to realize that criteria exist for evaluating estimators. One estimator may be preferred in one situation and a second competing estimator may be preferred in another situation. There is a tendency to use estimators whose good properties depend on the data following a normal distribution, even though the data are not normally distributed. This is unfortunate, since more appropriate methods of analysis are available.

An estimator that usually has good properties whatever the situation is called **robust.** The sample mean is a robust estimator of the population mean.


### MAXIMUM LIKELIHOOD

Having accepted the fact that estimators should have certain desirable properties, how do we find an estimator with such properties in the first place? Again there are many ways of doing this, and no single approach is preferable in every situation. A full discussion of these methods requires mathematical details that are beyond the scope of this book, but we must describe two very important approaches to deriving estimators, since they are frequently mentioned in the literature. One is based on a method known as maximum likelihood estimation and the other as least squares estimation. The method of least squares will be discussed in Chapter 9. **Maximum likelihood estimates** of parameters such as the population mean and variance are those values of the parameters that make the probability, or likelihood, of our sample as large as possible—a maximum. Intuitively, they are those parameter values that make the data we observe "most likely" to occur in the sample.

A simple example will illustrate the principle. Suppose we wish to estimate the proportion of a population that is male. We take a random sample of n persons from the population and observe in the sample y male and n-y female. Suppose now the true population proportion is $\pi$, so that each person in the sample has a probability $\pi$ of being male. What then is the probability, or likelihood, of our sample containing y male and n-y female? We learned in Chapter 5 that the probability of this happening is given by the binomial distribution, with parameters n and $\pi$; i.e.,

$$P(y \text{ male}) = \frac{n!}{y!(n-y)!} \, \pi^y \, (1-\pi)^{n-y}$$

Now we ask: What value of $\pi$ makes this probability the largest—a maximum? We shall call that value the maximum likelihood estimate of $\pi$. We shall not prove it here, but it can be shown that in this case the likelihood is largest when we set $\pi$ equal to y/n, the sample proportion of male persons. [It is instructive to take the time to verify this numerically. Suppose, for example, n = 5 and y = 2. Calculate P(y male) for various values of $\pi$ (e.g., $\pi$ = 0.2, 0.3, 0.4, 0.5). You will find that it is largest when $\pi$ = 0.4.] Thus, the maximum likelihood estimate of $\pi$ in this situation is y/n. Analogously, the maximum likelihood estimator of $\pi$ is Y/n, where Y is the random variable denoting the number of male persons we find in samples of size n from the population.

Except as noted in Appendix 1, maximum likelihood estimators have the following important properties for samples comprising a very large number of study units; i.e., as the number of study units tends to infinity, maximum likelihood estimators have the following (so-called **asymptotic**) properties

1. They are unbiased: the mean of the estimator (i.e., the mean of many, many estimates) will equal the true value of the parameter being estimated.
2. They are efficient: the variance of the estimator (i.e., the vari-

ance of many, many estimates) is the smallest possible for any asymptotically unbiased estimator.

3. The estimators are normally distributed. The utility of this last property will become apparent later in this chapter.

In certain instances, maximum likelihood estimators have some of these properties for all sample sizes, n; in general, however, they have these properties only for very large sample sizes. How large n must be for these properties to be enjoyed, at least approximately, depends on the particular situation, and is often unknown. Nevertheless, the principle of maximum likelihood estimation has a great deal of intuitive appeal and it is widely used.

## ESTIMATING INTERVALS

In the literature, estimates are often given in the form of a number plus or minus another number. For example, the mean serum cholesterol level for a group of 100 medical students might be reported as 186 ± 32 mg/100 ml. Unfortunately, these numbers are often reported without any explanation of what the investigator is attempting to estimate. It is a standard convention for the first number (186) to be the parameter estimate, but there is no standard convention regarding what the number after the ± sign represents. In some instances, the second number is a simple multiple (usually 1 or 2) of the estimated standard deviation of the random variable of interest. In other instances, it may be some multiple of the standard error of the estimator. In our example, 186 ± 32 mg/100 ml represents the mean serum cholesterol level plus or minus one standard deviation, estimated from a group of 100 medical students. If we divide 32 mg/100 ml by the square root of the sample size (i.e., by 10), we obtain the standard error of the mean, equal to 3.2 mg/100 ml. Thus the interval 186 ± 3.2 mg/100 ml represents the mean serum cholesterol level plus or minus one standard error of the mean. Clearly it is important to know which is being quoted, and to know whether the number after the ± sign is once, twice or some other multiple of the estimated standard deviation or standard error of the mean. We now turn to a discussion of how the different intervals should be interpreted.

We shall see later in this chapter that the standard error of the mean is used to define an interval that we believe contains the true value of the mean. For now, we shall suppose we wish to estimate an interval that includes most of the population values. Suppose that our population of cholesterol values is normally distributed with mean $\mu$ and standard deviation $\sigma$. Then we know that approximately two thirds of the population lies between $\mu - \sigma$ and $\mu + \sigma$, i.e., in the interval $\mu \pm \sigma$. We can estimate this interval by $\bar{y} \pm s$ (i.e., $186 \pm 32$ mg/100 ml). Thus, provided cholesterol levels are normally distributed and our sample size is large enough for the estimates to be close to the true parameters, we can expect about two thirds of the population to have cholesterol levels between 154 and 218 mg/100 ml. Similarly, we know that about 95% of the population lies in the interval $\mu \pm 2\sigma$, which for our example is estimated by $186 \pm 64$. Thus we might estimate that 95% of the population cholesterol levels are between 122 and 250 mg/100 ml.

Such intervals, or limits, are often quoted in the literature in an attempt to define a "**normal range**" for some quantity. They must, however, be very cautiously interpreted on two counts. First, the assumption of normality is not a trivial one. Suppose it was quoted, on the basis of the data in Table 4.1, that $\bar{y} \pm 2s$ for triglyceride levels in male medical students is $111 \pm 118$ mg/100 ml ($y = 111$, $s = 59$). Should we expect 95% of the population to be in this interval? Note that this interval is from $-7$ to 229 mg/100 ml. Triglyceride levels are not at all normally distributed, but follow a very positively skewed distribution in the population (see Figures 3.2 and 3.4); this explains why $\bar{y} - 2s$ is negative and hence an impossible quantity. We must not assume that $\mu \pm 2\sigma$ includes about 95% of the population for a non-normally distributed random variable. Second, there is an implicit assumption (in addition to normality) that the sample estimates are close to the true parameter values. It is rarely realized that it takes very large sample sizes for this assumption to be adequately met. If the sample size is less than 50, then $\bar{y} \pm 2s$ includes, on an average, less than 95% of a normally distributed population. When the sample size is 10, it includes on an average only 92%, and when the sample size is down to 8, it includes on an average less than 90%. Although these consequences might be considered small enough to neglect, it must be realized that these percentages are only averages.

**TABLE 6.1.** Approximate Proportion of Samples from a Normal Distribution in Which the Estimated Mean ± 2 s.d. Will Fail to Include the Indicated Percentage of the Population

| | Percentage of the Population | | |
|---|---|---|---|
| Sample Size | 75 | 90 | 95 |
| 10 | 0.12 | 0.18 | 0.24 |
| 20 | 0.09 | 0.12 | 0.15 |
| 30 | 0.08 | 0.10 | 0.12 |
| 40 | 0.07 | 0.09 | 0.11 |
| 50 | 0.07 | 0.09 | 0.11 |
| 100 | 0.06 | 0.07 | 0.08 |

The results from any one sample could well include less than 75% of the population, as can be seen from Table 6.1. We see from this table, for example, that when the sample size is 10 there is a 0.12 probability that $\bar{y} \pm 2s$ includes less than 75% of the population. Even when the sample size is 100 there is still a 0.06 probability that $\bar{y} \pm 2s$ includes less than 75% of the population. Thus, if cholesterol levels are normally distributed, the interval 122–250 mg/100ml has a small but non-negligible probability (0.06) of including less than 75% of the population.

## THE STANDARD NORMAL (GAUSSIAN) DISTRIBUTION AND THE DISTRIBUTION OF THE SAMPLE MEAN

Suppose a random variable Y is normally distributed in the population with mean $\mu$ and standard deviation $\sigma$. We have seen that about 68% of the population then lies between $\mu - \sigma$ and $\mu + \sigma$, and about 95% of the population lies between $\mu - 2\sigma$ and $\mu + 2\sigma$.

Now let us subtract the mean from Y and divide the difference by the standard deviation. Call the result Z, i.e.,

$$Z = \frac{Y - \mu}{\sigma}.$$

Then Z is also normally distributed, but has mean 0 and standard deviation 1. This is pictured in Figure 6.1. The distribution of Z is called the "**standard normal distribution.**" Any random variable that has mean 0 and standard deviation 1 is said to be "standardized."

Distribution of Z

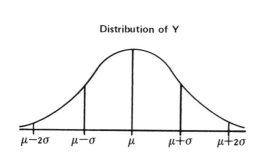

Distribution of Y

$\mu-2\sigma$    $\mu-\sigma$    $\mu$    $\mu+\sigma$    $\mu+2\sigma$    $-2$    $-1$    $0$    $1$    $2$

**FIGURE 6.1** Normal density function of the random variable Y, with mean $\mu$ and standard deviation $\sigma$, and of the standardized random variable Z, with mean 0 and standard deviation 1.

(The symbol Z, or the term "Z-score," is often used in the literature to indicate a variable that has been standardized—sometimes in an unspecified manner.) For a standardized **normal** random variable, about 68% of the population lies between $-1$ and $+1$, and about 95% of the population lies between $-2$ and $+2$. (More precisely, 95% of the population lies between $-1.96$ and $+1.96$; we shall continue to use the approximate values $\pm 2$).

Mathematically, if Y is normally distributed, we can write

$$P(\mu - 2\sigma \le Y \le \mu + 2\sigma) = 0.95$$

and

$$P\left(-2 \le \frac{Y - \mu}{\sigma} \le 2\right) = 0.95$$

The 97.5th percentile of the standard normal distribution is about 2, and the 2.5th percentile is about $-2$.

Now consider $\overline{Y}$, the mean of n such normally distributed random variables. We have already learned that $\overline{Y}$ has mean $\mu$ and standard deviation $\sigma/\sqrt{n}$. Furthermore, it is normally distributed. As an example, suppose we have a population in which the mean height is 67 inches and the standard deviation is 3 inches. We take samples of four persons each and average their heights. Then the distribution of these averages is normal with mean 67 inches and standard deviation $3/\sqrt{4}$ inches = 1.5 inches. This is pictured in Figure 6.2. Notice that whereas about 95% of the population lies in the interval 61 to 73 inches, about 95% of the means lie in the interval 64 to 70 inches.

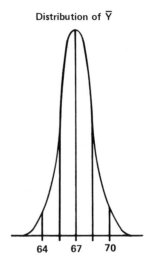

Distribution of $\overline{Y}$

Distribution of Y

FIGURE 6.2 Normal density function of a person's height Y, with mean 67 inches and standard deviation 3 inches, and of the average of four persons' heights Y, with same mean 67 inches but standard deviation 1.5 inches.

Now let us subtract the mean from $\overline{Y}$ and divide the difference by *its* standard deviation, i.e.,

$$Z = \frac{\overline{Y} - \mu}{\sigma_{\overline{Y}}}.$$

We denote the result Z, since it is the letter usually used for a standardized normal random variable; with about 95% probability it lies between $-2$ and $+2$. Thus we can write

$$P(-2 \leq Z \leq 2) \doteq 0.95$$

$$\text{or } P\left(-2 \leq \frac{\overline{Y} - \mu}{\sigma_{\overline{Y}}} \leq 2\right) = 0.95$$

which is equivalent to

$$P(\overline{Y} - 2\sigma_{\overline{Y}} \leq \mu \leq \overline{Y} + 2\sigma_{\overline{Y}}) \doteq 0.95$$

(The derivation of this equivalence is given in Appendix 1.) Remember, $\sigma_{\overline{Y}} = \sigma_Y / \sqrt{n}$ for a sample of size n.

Let us now summarize the various facts we have learned about the sample mean, $\overline{Y}$:

1. The mean of $\overline{Y}$ is $\mu$.
2. The variance of $\overline{Y}$ is $\sigma^2/n$ (i.e., $\sigma_{\overline{Y}}^2 = \sigma_Y^2/n$).
3. The standard deviation of $\overline{Y}$ is $\sigma/\sqrt{n}$ (i.e., $\sigma_Y = \sigma_{\overline{Y}}/\sqrt{n}$).
4. $\overline{Y}$ is normally distributed. This is strictly true only if Y is normally distributed. It is a remarkable fact, however, that it also tends to be true, for moderately large or large sample sizes, almost irrespective of how Y is distributed (the word "almost" is added to allow for some special situations that, although mathematically possible, do not usually occur in practice). Usually, a mean of five or more observations is for all intents and purposes normally distributed.
5. It therefore follows, provided Y is based on five or more observations, that

$$P(\overline{Y} - 2\sigma_Y/\sqrt{n} \leq \mu \leq \overline{Y} + 2\sigma_Y/\sqrt{n}) \doteq 0.95.$$

### CONFIDENCE LIMITS

Consider once more the triglyceride levels in Table 4.1, for which $\overline{y}$ = 111 mg/100 ml, s = 59 mg/100 ml, and n = 30. The standard error of the mean is $59/\sqrt{30}$ = 11 mg/100 ml. Let us assume for the moment that the true population values are $\sigma_Y$ = 59 and $\sigma_{\overline{Y}}$ = 11 mg/100 ml. (We shall see how to avoid this assumption later.) Thus the interval $\overline{y} \pm 2\sigma_{\overline{Y}}$ is 111 $\pm$ 22, or 89 to 133 mg/100 ml. Can we therefore say that there is about a 95% probability that $\mu$ lies in this interval? In other words, can we write

$$P(89 \leq \mu \leq 133) \doteq 0.95 ?$$

Regardless of how good the approximation might be, such an expression is impossible. The true mean $\mu$ is a fixed quantity, not a random variable, and the fact that we do not know it does not alter this. Similarly, 89 and 133 are fixed quantities. Either $\mu$ lies between 89 and 133 mg/100 ml, or it does not. We cannot talk about the "probability" of this being the case, since probability is a property associated with random variables. Thus it is meaningful to write

$$P(\overline{Y} - 2\sigma_{\overline{Y}} \leq \mu \leq \overline{Y} + 2\sigma_{\overline{Y}}) \doteq 0.95$$

but not to write

$$P(\bar{y} - 2\sigma_{\bar{Y}} \leq \mu \leq \bar{y} + 2\sigma_{\bar{Y}}) \doteq 0.95.$$

Now you can see why we have taken pains to distinguish between $\bar{Y}$ and $\bar{y}$. The former is a random variable, about which we can make a probability statement, the latter is a fixed quantity about which we cannot make a probability statement.

Despite the fact that we cannot talk about the probability of $\mu$ lying between 89 and 133 mg/100 ml, this interval is clearly somehow related to the unknown mean $\mu$, and we should be reasonably certain (perhaps "95% certain") that $\mu$ is in fact between 89 and 133 mg/100 ml. We call these numbers the 95% **confidence limits**, or **confidence interval**, for $\mu$, and we say that $\mu$ lies between 89 and 133 mg/100 ml with 95% **confidence.**

Thus $\bar{y} \pm 2\sigma_{\bar{Y}}$ (i.e., $\bar{y} \pm 2\sigma_Y/\sqrt{n}$) gives an approximate 95% confidence interval for the mean. This is strictly true only if Y is normally distributed, but tends to be true, for moderately large or large samples, whatever the distribution of Y. A confidence interval is to be interpreted as follows: if we were to find many such intervals, each from a different sample but in exactly the same fashion, then in the long run about 95% of our intervals would include the true mean and 5% would not. We cannot say that there is a 95% probability that the true mean lies between the two values we obtain from a particular sample, but we can say that we have 95% confidence that it does so. We shall now briefly indicate how confidence limits are calculated in several specific situations.

## CONFIDENCE LIMITS FOR A PROPORTION

Suppose we wish to estimate the probability a newborn is male. We take a random sample of 1,000 births and find 526 are male. We therefore estimate the proportion of births that are male as being 0.495. What would the 95% confidence limits be? Since 0.495 is a maximum likelihood estimate based on a large sample, it can be considered as the outcome of a normally distributed random variable. Now if Y follows a binomial distribution with parameters n and $\pi$, then it can be shown that the variance of Y/n (y/n is 526/1,000 in our particular sample) is $\pi(1 - \pi)/n$. Thus Y/n is about normally distributed with mean $\pi$ and standard deviation $\sqrt{\pi(1 - \pi)/n}$, and so an approximate 95% confidence interval for $\pi$ is

$$y/n \pm 2\sqrt{\pi(1 - \pi)/n}$$

which we estimate by

$$0.526 \pm 2\sqrt{0.526(1 - 0.526)/1000} = 0.526 \pm 0.031.$$

Thus we have about 95% confidence that the true proportion lies between 0.494 and 0.557.

Notice that to calculate this interval we substituted our estimate, 0.526, for $\pi$. For a large sample this is adequate. Had the sample been small, we should not assume $Y/n$ is normally distributed and it would have been necessary to use special tables that have been calculated from the binomial distribution. As a rule of thumb, the approximation is adequate provided both $n\pi$ and $n(1 - \pi)$ are greater than 5. Note also that since the estimator is about normally distributed, in large samples a 95% confidence interval can always be obtained from a maximum likelihood estimate by adding and subtracting twice the standard error of the estimator.

### CONFIDENCE LIMITS FOR A MEAN

Consider our example in which we determined 95% confidence limits for mean triglyceride level as being 89 to 133 mg/100 ml. Recall that we assumed we knew $\sigma_Y = 59$ and $\sigma_{\bar{Y}} = 11$ mg/100 ml, whereas in fact these were really the sample estimates $s_Y$ and $s_{\bar{Y}}$, respectively. Usually we do not know the true standard deviation, $\sigma_Y$, so to be of any practical use we must be able to calculate confidence limits without it. Does it make any difference if we simply substitute $s_{\bar{Y}}$ for $\sigma_{\bar{Y}}$? The answer is that it does not, for all practical purposes, if the sample size is more than 30. When the sample size is smaller than this we must allow for the fact that we do not know $\sigma_Y$, as follows.

We have seen that if $Y$ is normally distributed, then $(\bar{Y} - \mu)/\sigma_{\bar{Y}}$ follows a standard normal distribution for which the approximate 2.5th and 97.5th percentiles are, respectively, $-2$ and $+2$. Analogously, substituting $S_{\bar{Y}}$ for $\sigma_{\bar{Y}}$, $(\bar{Y} - \mu)/S_{\bar{Y}}$ follows a distribution called **Student's t distribution with n $-$ 1 degrees of freedom**. Like the standard normal distribution, this is symmetric about zero, so that the

2.5th percentile is simply the negative of the 97.5th percentile. Denote the 97.5th percentile $t_{97.5}$. Then the 95% confidence limits are $\bar{y} \pm t_{97.5}s_{\bar{y}}$.

The distribution of $(\bar{Y} - \mu)/S_{\bar{y}}$ was derived by a mathematician who worked for the Guinness Brewing Company in Ireland and who published his statistical papers under the pseudonym "Student." This quantity is denoted t, and hence the name "t distribution." Just as the normal distribution is a family of distributions, a particular one being determined by the parameters $\mu$ and $\sigma$, so is the t distribution a family of distributions—but in this case the particular distribution is determined by a parameter known as the "number of degrees of freedom," a concept we shall explain shortly. Each t distribution is similar to the standard normal distribution but has thicker tails, as illustrated in Figure 6.3. The fewer the degrees of freedom, the thicker the tails are. As the number of degrees of freedom becomes indefinitely large (in which case $\sigma = s$), the t distribution becomes the same as the standard normal distribution. This can be seen in the following 97.5th percentiles, abstracted from Table 1 in Appendix 2:

$$
\begin{array}{ll}
\text{1 degree of freedom} : & 12.706 \\
\text{10} \text{---------------------} : & 2.228 \\
\text{30} \text{---------------------} : & 2.042 \\
\text{(Standard normal) } \infty \text{---------------------} : & 1.960
\end{array}
$$

Now you can see the basis for saying that for all practical purposes it makes no difference if we substitute $s_{\bar{y}}$ for $\sigma_{\bar{y}}$ when the sample size is

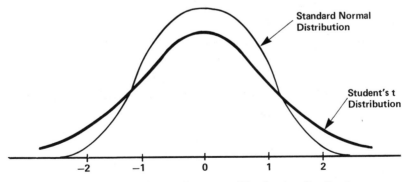

**FIGURE 6.3** Example of the density function of Student's t distribution compared to that of the standard normal distribution.

more than 30; 2 is almost as close an approximation to 2.042 ($t_{97.5}$ when there are 30 degrees of freedom, corresponding to n = 31) as it is to 1.960 (the 97.5th percentile of the standard normal distribution).

Now let us go back once more to our example of triglyceride levels (from the data in Table 4.1), for which y = 111 mg/100 ml, $s_{\bar{y}}$ = 59 mg/100 ml, and n = 30. Calculating with more accuracy than previously, we find $s_{\bar{y}}$ = 59/√30 = 10.77 mg/100 ml, and from Table 1 in Appendix 2 we find that for n − 1 = 29 degrees of freedom, $t_{97.5}$ = 2.045. From this we find that $\bar{y} \pm t_{97.5}s_{\bar{y}}$ is 111 ± (2.045)(10.77) = 111 ± 22.0 mg/100 ml. Thus in this case our earlier approximation was adequate. Had the sample size been much smaller (e.g., 10 or less), however, the approximation $\bar{y} + 2s_{\bar{y}}$ would have led to an interval that is much too short.

## THE CONCEPT OF DEGREES OF FREEDOM

The term **degrees of freedom**, abbreviated d.f., will occur again and again, not only in connection with the t distribution, but also in many other problems. Basically, the degrees of freedom refer to the number of "independent" observations in a quantity. You know that the sum of the angles in a triangle is equal to 180°. Suppose you were asked to choose the three angles of a triangle. You have two degrees of freedom in the sense that you may choose two of the angles, but then the other is automatically determined because of the restriction that the sum of the three is 180°. Suppose you are asked to choose three numbers with no restrictions on them. You have complete freedom of choice in specifying all three numbers and hence, in that case, you have three degrees of freedom.

Now suppose you are asked to choose six numbers (which we shall call $y_1$, $y_2$, $y_3$, $y_4$, $y_5$ and $y_6$) such that their sum is 40 and also such that the sum of the first two is 16. There are six numbers to be specified, but you do not have freedom of choice for all six. You have to take into account the restrictions

$$y_1 + y_2 = 16$$

and
$$y_1 + y_2 + y_3 + y_4 + y_5 + y_6 = 40.$$

As soon as you select $y_1$, then $y_2 = 16 - y_1$, and so $y_2$ is completely determined. Of the remaining numbers, $y_3 + y_4 + y_5 + y_6 = 40 - 16 = 24$. Thus only three of the numbers $y_3$, $y_4$, $y_5$, and $y_6$ can be freely chosen. If we choose $y_3$, $y_4$, and $y_5$, for example, $y_6$ is predetermined as follows

$$y_6 = 24 - (y_3 + y_4 + y_5).$$

Hence, the total number of degrees of freedom in this example is $1 + 3 = 4$.

In computing a variance, we use as the divisor a number that makes the variance an unbiased estimator of the population variance. This divisor is the number of degrees of freedom associated with the estimator once an estimate of the mean has been made. Recall that our divisor for the variance is the size of the sample minus one (i.e., n − 1). Once the mean is fixed at its sample value, there are only n − 1 degrees of freedom associated with permissible values of the numbers used to compute the variance. This same number is also the number of degrees of freedom associated with the estimated standard deviation, and with the t-distribution used to obtain confidence limits for the mean from the formula $\bar{y} \pm t s_{\bar{y}}$.

Other types of problems with a variety of restrictions and degrees of freedom are considered in subsequent chapters of this book. In every instance the number of degrees of freedom is associated with a particular statistic (such as $s_Y$). It is also the appropriate value to use as the parameter of a distribution (such as the t-distribution) when using that statistic for a particular purpose (such as calculating confidence limits).

## CONFIDENCE LIMITS FOR THE DIFFERENCE BETWEEN TWO MEANS

Suppose we compare two drugs, A and B, each aimed at lowering serum cholesterol levels. Drug A is administered to one group of patients (sample 1) and drug B to a second group (sample 2), with the patients randomly assigned to the two groups so that the samples are independent. If we use the estimators $\bar{Y}_1$ and $\bar{Y}_2$ to find estimates $\bar{y}_1$ and $\bar{y}_2$ of the post-treatment serum cholesterol means for drug A and drug B, respectively, we might want to construct a confidence interval for the true difference $\mu_1 - \mu_2$. In this situation a 95% confidence interval would be given by

$$\bar{y}_1 - \bar{y}_2 \pm t_{97.5}s_{\bar{Y}_1-\bar{Y}_2}$$

where $t_{97.5}$ is the 97.5th percentile of the t distribution, with degrees of freedom equal to that associated with $s_{\bar{Y}_1-\bar{Y}_2}$, the standard error of $\bar{Y}_1 - \bar{Y}_2$. If the two samples have the same true variance, $\sigma^2$, then the respective sample variances, $s_1^2$ and $s_2^2$, are both estimates of the same quantity, $\sigma^2$. In such instances we can average or "pool" the sample variances to obtain a **pooled estimate**, $s_p^2$, of $\sigma^2$:

$$s_p^2 = \frac{(n_1 - 1)s_1^2 + (n_2 - 1)s_2^2}{n_1 + n_2 - 2}$$

Note that when we pool $s_1^2$ and $s_2^2$, we weight each by the number of degrees of freedom associated with it. Note also that we obtain a (weighted) average of sample **variances**, not of standard deviations. We take the square root of this pooled variance to obtain the (sample) pooled standard deviation, $s_p$. Next, you need to know that when we have two **independent** random variables, the variance of their difference is equal to the sum of their variances. Thus the variance of $\bar{Y}_1 - \bar{Y}_2$ is the variance of $\bar{Y}_1$ plus the variance of $\bar{Y}_2$, i.e.,

$$\sigma^2_{\bar{Y}_1-\bar{Y}_2} = \frac{\sigma^2}{n_1} + \frac{\sigma^2}{n_2} = \sigma^2\left(\frac{1}{n_1} + \frac{1}{n_2}\right)$$

and it follows that the standard deviation of $\bar{Y}_1 - \bar{Y}_2$ is the square root of this, or $\sigma\sqrt{1/n_1 + 1/n_2}$. The standard error of $\bar{Y}_1 - \bar{Y}_2$ is obtained by substituting the estimate $s_p$ for $\sigma$ in this expression, so that the confidence interval is

$$\bar{y}_1 - \bar{y}_2 \pm t_{97.5}s_{\bar{Y}_1-\bar{Y}_2} = \bar{y}_1 - \bar{y}_2 \pm t_{97.5}s_p\sqrt{\frac{1}{n_1} + \frac{1}{n_2}},$$

in which $t_{97.5}$ is the 97.5th percentile of the t distribution, with $n_1 + n_2 - 2$ degrees of freedom. The number of degrees of freedom associated with $s_p$ is the sum of the number of degrees of freedom associated with $s_1$ and the number associated with $s_2$.

If the two samples do not have the same true variance, other methods, which we shall not detail here, must be used. The same method, however, leads to a good approximation of a 95% confidence interval even if the true variances are unequal, provided $n_1 = n_2$ (i.e., provided the two samples have the same size). When the two samples have different sizes we must first determine whether it is reasonable to suppose that the two true variances are equal. This, and other similar topics, are the subject of the next chapter.

## SUMMARY

1. An estimator is a rule for calculating an estimate from a set of sample values. It is a random variable that takes on different values—estimates—from sample to sample. The mean of n independent random variables, each with variance $\sigma^2$, has variance $\sigma^2/n$ and standard deviation $\sigma/\sqrt{n}$. The latter, or its estimate $s/\sqrt{n}$, is called the standard error of the mean.

2. An unbiased estimator is one whose mean is equal to the parameter being estimated. The mean and variance of a random sample are unbiased estimators of the population mean and variance. An efficient estimator is an unbiased estimator that has minimum variance. In the case of normally, binomially, or Poisson-distributed random variables, the sample mean is a minimum variance unbiased estimator. The sample mean is a robust estimator (i.e., a good estimator of the population mean in a wide variety of situations).

3. Maximum likelihood estimates are parameter values that make the likelihood (probability density) of the sample a maximum. In large samples, maximum likelihood estimators are usually unbiased, efficient, and normally distributed. Sometimes they also have these properties in small samples.

4. The estimated mean ± 2 standard deviations is often calculated as a "normal range" that contains about 95% of the population values. There is no guarantee that this approximation is good unless the population is normally distributed and the interval is calculated from a large sample.

5. A standardized random variable has mean 0 and standard deviation 1. It is obtained by subtracting from a random variable its mean and dividing the result by its standard deviation. For the standard normal distribution, about 68% of the population lies within one standard deviation and 95% within two standard deviations from the mean. Sample means tend to be normally distributed regardless of the form of the distribution being sampled.

6. A parameter can be said to lie within a specified interval with a

certain degree of confidence, not with any degree of probability. The estimator, but not the estimate, of a 95% confidence interval can be said to have 95% probability of including the parameter. A particular 95% confidence interval for the mean should be interpreted as follows: if many such intervals were to be calculated in the same fashion, each from a different sample, then in the long run 95% of such intervals would include the true mean.

7. For large samples, the maximum likelihood estimate of a parameter $\pm$ twice the standard error is an approximate 95% confidence interval for that parameter. For sample sizes larger than 30, the sample mean $\pm$ two standard errors of the mean provide approximate 95% confidence limits for the mean. For smaller samples and a normally distributed random variable, 95% confidence limits are given by $\bar{y} \pm t_{97.5}s/\sqrt{n}$, where $t_{97.5}$ is the 97.5th percentile of Student's t distribution with $n-1$ degrees of freedom.

8. If two independent samples come from populations with the same common variance, a pooled estimate, $s_p^2$, of the variance can be obtained by taking a weighted average of the two sample variances, $s_1^2$ with $n_1 - 1$ degrees of freedom and $s_2^2$ with $n_2 - 1$ degrees of freedom, weighting by the number of degrees of freedom. The pooled estimate then has $n_1 + n_2 - 2$ degrees of freedom and can be used to determine a 95% confidence interval for the difference between the two means: $\bar{y}_1 - \bar{y}_2 \pm t_{97.5}s_p\sqrt{1/n_1 \ 1/n_2}$. This same interval is about correct even if the two variances are different, provided the two sample sizes $n_1$ and $n_2$ are equal.

## PROBLEMS

1. An unbiased estimate
A. is equal to the true parameter
B. has the smallest variance of all possible estimates
C. is never an efficient estimate
D. has mean equal to the true parameter
E. is always a maximum likelihood estimate

2. If the standard error of the mean obtained from a sample of nine observations is quoted as being three units, then nine units is
A. the true variance of the population

B. the estimated variance of the population
C. the true standard deviation of the population
D. the estimated standard deviation of the population
E. none of the above


3. An experimenter reports that on the basis of a sample of size 10, he calculates the 95% confidence limits for mean height to be 66 and 74 inches. Assuming his calculations are correct, this result is to be interpreted as meaning
   A. there is a 95% probability that the population mean height lies between 66 and 74 inches
   B. we have 95% confidence that a person's height lies between 66 and 74 inches
   C. we have 95% confidence that the population mean height lies between 66 and 74 inches
   D. 95% of the population has a height between 66 and 74 inches
   E. none of the above


4. A 99% confidence interval for a mean
   A. is wider than a 95% confidence interval
   B. is narrower than a 95% confidence interval
   C. includes the mean with 99% probability
   D. excludes the mean with 99% probability
   E. is obtained as the sample average plus two standard deviations


5. In a series of journal articles, investigator A reported his data, which are about normally distributed, in terms of a mean plus or minus two standard deviations, while investigator B reported his data in terms of a mean plus or minus two standard errors of the mean. The difference between the two methods is
   A. investigator A is estimating the extreme percentiles whereas investigator B is estimating the most usual percentiles
   B. investigator A is estimating the range that he thinks contains 95% of the means, whereas investigator B is estimating the range that he thinks contains 95% of the medians
   C. investigator A is estimating the range that he thinks contains about 95% of his data values, whereas investigator B is esti-

mating the range that he thinks (with 95% confidence) contains the true mean being estimated

D. investigators A and B are really estimating the same range, but are just using different systems of reporting

E. none of the above

6. A 95% confidence interval implies that
A. the t-test gives correct intervals 95% of the time
B. if we repeatedly select representative samples and construct such interval estimates, 95 out of 100 of the intervals would be expected to bracket the true mean
C. the hypothesis will be false in 95 out of 100 such intervals
D. the probability that the interval is false is 95%
E. there is a 95% probability that the underlying distribution is normal

7. Student's t distribution with one degree of freedom
(1) has fatter tails than a standard normal distribution
(2) can be used to obtain confidence limits for the mean of a normal distribution from a sample of two observations
(3) has mean 0
(4) has variance 1
  A. 1, 2, and 3
  B. 1 and 3
  C. 2 and 4
  D. only 4
  E. 1, 2, 3, and 4

8. In a sample of 100 normal women between the ages of 25 and 29 years, systolic blood pressure was found to follow a normal distribution. If the sample mean pressure was 120 mmHg and the standard deviation was 10 mmHg, what interval of blood pressure would represent an approximate 95% confidence interval for the true mean?
  A. 118 to 122 mmHg.
  B. 100 to 140 mmHg.
  C. 119 to 121 mmHg.

D. 110 to 130 mmHg.

E.   90 to 150 mmHg.

9. A researcher is interested in the population variability of a normally distributed trait, and finds two estimates of its standard deviation in the literature. These two estimates are similar, and the researcher wishes to average them to obtain one overall estimate. The best procedure is to

A. take the simple average of the estimated standard deviations

B. take a weighted average of the estimated standard deviations, weighting them by their degrees of freedom

C. take a simple average of the squares of the estimated standard deviations, and then take the square root of the result

D. take a weighted average of the squares of the estimated standard deviations, weighting them by their degrees of freedom, and then take the square root of the result

E. none of the above

10. An investigator is interested in the mean cholesterol level of patients with myocardial infarction. On the basis of a random sample of 50 such patients, a 95% confidence interval for the mean has a width of 10 mg/100 ml. How large a sample would be expected to have given an interval with a width of about 5 mg/100 ml?

A. 100

B. 200

C. 300

D. 400

E. 800

## CHAPTER 7

# Tests of Hypotheses and Significance Levels

**Key Concepts:** null hypothesis, test criterion, significance level, p-value; one-sided (one-tail), two-sided (two-tail) tests; F distribution; two-sample t-test, paired t-test; distribution-free methods: rank sum test, signed rank sum test, sign test; Type I error, validity; Type II error, power.

## PRINCIPLE OF SIGNIFICANCE TESTING

A hypothesis is a contention that may or may not be true, but is provisionally assumed to be true until new evidence suggests otherwise. A hypothesis may be proposed from a hunch, from a guess, or on the basis of preliminary observations. A statistical hypothesis is a contention about a population, and we investigate it by performing a study on a sample collected from that population. We then examine the sample information to see how consistent the "data" are with the hypothesis under question; if there are discrepancies, we tend to disbelieve the hypothesis and reject it. So the question arises, how inconsistent with the hypothesis do the sample data have to be before we are prepared to reject the hypothesis? It is to answer questions such as this that we use statistical tests of hypotheses, or **significance** tests. In general, three steps are taken in performing a significance test:

1. Convert the **research hypothesis** to be investigated into a specific statistical **null hypothesis.** The null hypothesis is a specific hypothesis that we try to disprove. It is usually expressed in terms of population parameters. For example, suppose that our

research hypothesis is that a particular drug will lower blood pressure. We randomly assign patients to two groups: group 1 to receive the drug and group 2 to act as controls without the drug. Our research hypothesis is that after treatment, the mean blood pressure of group 1, $\mu_1$, will be less than that of group 2, $\mu_2$. In this situation the specific null hypothesis that we try to disprove is $\mu_1 = \mu_2$. In another situation we might disbelieve a claim that a new surgical procedure will cure at least 60 % of cases of a particular type of cancer, and our research hypothesis would be that the probability of cure, $\pi$, is less than this. The null hypothesis that we would try to disprove is $\pi = 0.6$. Another null hypothesis might be that two variances are equal (i.e., $\sigma_1^2 = \sigma_2^2$). Notice that these null hypotheses can be expressed as a function of parameters equalling zero, and hence the terminology **null** hypothesis

$$\mu_1 - \mu_2 = 0, \ \pi - 0.6 = 0, \ \sigma_1^2 - \sigma_2^2 = 0.$$

2. Decide on an appropriate test criterion to be calculated from the sample values. We view this calculated quantity as one particular value of a random variable that takes on different values in different samples. Our statistical test will utilize the fact that we know how this quantity is distributed from sample to sample if the null hypothesis is true. This distribution is the sampling distribution of the test criterion **under** the null hypothesis, and is often referred to as the null distribution. We shall give examples of several test criteria later in this chapter.

3. Calculate the test criterion from the sample and determine the corresponding significance level. The test criterion is compared with its sampling distribution to quantify how "likely" it is under the null hypothesis. If it is likely, we conclude that the evidence is insufficient to reject the null hypothesis; for the time being, we accept the null hypothesis. If, on the other hand, the test criterion is unlikely, we reject the null hypothesis. The **significance level** will become clearer when we consider some examples. It is a probability that is at least as large as the probability of the observed or any more extreme sample occurring, if the null hypothesis is true.

# TESTING A HYPOTHESIS ABOUT A POPULATION MEAN

As an example of a particular significance test, suppose our research hypothesis is that the mean weight of adult male patients who have been on a weight reduction program is less than 200 lb. We wish to determine whether this is so. The three steps are as follows:

1. We try to disprove that the mean weight is 200 lb. The null hypothesis we take is therefore that the mean weight is 200 lb. We can express this null hypothesis as $\mu = 200$.
2. Suppose now that, among male patients who have been on the program, weight is normally distributed with mean 200 lb. We weigh a random sample of n such men and calculate the sample mean $\bar{y}$ and standard deviation $s_Y$. Then we know that

$$\frac{\bar{Y} - \mu}{s_{\bar{Y}}} = \frac{\bar{Y} - 200}{s_{\bar{Y}}} \text{ , where } s_{\bar{Y}} = s_Y/\sqrt{n} \text{ ,}$$

follows Student's t distribution with n − 1 degrees of freedom. We therefore use

$$t = \frac{\bar{y} - 200}{s_{\bar{Y}}}$$

as our test criterion: we know its distribution from sample to sample if the mean of Y is in fact 200 and Y is normally distributed.
3. Suppose our sample consisted of n = 10 men on the program, and we calculated from this sample $\bar{y} = 184$ and $s_Y = 26.5$ (and hence $s_{\bar{Y}} = 26.5/\sqrt{10} = 8.38$). Thus

$$t = \frac{184 - 200}{8.38} = -1.9 \text{ .}$$

We now quantify the "probability" of such a finding as follows. From a table of Student's t distribution (see Table 1, Appendix 2) we find, for 10 − 1 = 9 degrees of freedom, the following percentiles:

| %: | 2.5 | 5 | 95 | 97.5 |
|---|---|---|---|---|
| t-value: | −2.262 | −1.833 | 1.833 | 2.262 |

(The table gives us $t_{95} = -t_5 = 1.833$ and $t_{97.5} = -t_{2.5} = 2.262$.)
Thus the value we found, $-1.9$, lies between the 2.5th and the 5th
percentiles. Pictorially, the situation is as illustrated in Figure 7.1. If
the value of t had been somewhere near 0, this would be a "typical"
value and there would be no reason to suspect the null hypothesis.
The t value of $-1.9$ lies, however, below the 5th percentile (i.e., if
the null hypothesis is true, the probability is less than 0.05 that t
should be as far to the left as $-1.9$). We are therefore faced with
what the famous statistician Sir Ronald Fisher (1890–1962) called a
"logical disjunction": either the null hypothesis is not true, or, if it is
true, we have observed a rare event—one that has less than 5% probability of occurring simply by chance. We say that the result is **significant at the 5% level**, or that the mean is significantly less than 200 lb.
at the 5% level. Symbolically this is often written, with no further
explanation, $p < 0.05$. It is understood that p stands for the probability of observing what we actually did observe, or anything more extreme, if the null hypothesis is true. Thus, in our example, p is the
area to the left of $-1.9$ under the curve of a t distribution with 9
degrees of freedom. The area is about 4.5%, and this fact can be
expressed as $p \doteq 0.045$. We can say the result is significant at the
4.5% level, because 4.5% is at least as large as p. Similarly, we can
say that the result is significant at the 0.1 level, because 0.1 is at least
as large as p. In fact we can say that the result is significant at any
level greater than 4.5%, and not significant at any level less than
4.5%.

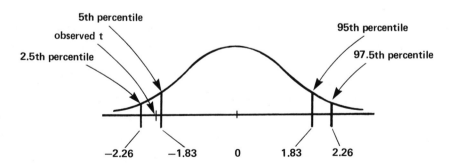

**FIGURE 7.1** Comparison of the observed $t = -1.9$ with Student's t distribution
with nine degrees of freedom.

However, most investigators arbitrarily pick a specific level (often 5% or 1%) as a cutoff, and decide to "reject" the null hypothesis for any result significant at that level. Notice, however, what the null hypothesis was (i.e., what was assumed to obtain the distribution of the test criterion). We assumed both a normal distribution and a mean of 200 lb. We also assumed, in order to arrive at a statistic that should theoretically follow the t distribution, a random sample of n weights. All these assumptions are part of the null hypothesis that is being tested, and departure from any one of them could be the cause of a significant result. Provided we do have a random sample and a normal distribution, however, either we have observed an unlikely outcome ($p \doteq 0.045$) or the mean is less than 200 lb.

It is a common (but arbitrary) convention to consider any value of p greater than 0.05 as "not significant." The idea behind this is that one should not place too much faith in a result that, by chance alone, would be expected to occur with a probability greater than 1 in 20. Other conventional phrases that are sometimes used are:

$$0.01 < p < 0.05: \quad \text{"significant."}$$
$$0.001 < p < 0.01: \quad \text{"highly significant."}$$
$$p < 0.001: \quad \text{"very highly significant."}$$

This convention is quite arbitrary, and arose in part because the cumulative probabilities of the various sampling distributions (such as the t distribution) are not easy to calculate, and so had to be tabulated. Typical tables, so as not to be too bulky, include just a few percentiles, such as the 90th, 95th, 97.5th, 99th, and 99.5th percentiles, corresponding to the tail probabilities 0.1, 0.05, 0.025, 0.01, and 0.005 for one tail of the distribution. Now that computers and calculators are commonplace, however, it is becoming more and more common to calculate and quote the actual value of p. It is clearly absurd to quote a result for which $p = 0.049$ as "significant" and one for which $p = 0.051$ as "not significant": it is far more informative to quote the actual p-values, which an intelligent reader can see are virtually identical in this case.

Note that we have defined the meaning of "significant" in terms of probability: in this sense it is a technical term, always to be interpreted in this precise way. This is often emphasized by saying, for example, "the result was statistically significant." Such a phrase, however, although making it clear that significance in a probability sense is meant, is completely meaningless unless the level of significance is also given. (The result of every experiment is statistically significant at the 100% level, because the significance level can be **any** probability at least as large as p!) It is important to realize that statistical significance is far different from biological significance. If we examine a large enough sample, even a biologically trivial difference can be made to be statistically significant. Conversely, a difference that is large enough to be of great biological significance can be statistically "not significant" if a very small sample size is used. We shall come back to this point at the end of the chapter.

Notice carefully the definition of p: the probability of observing what we actually did observe, **or anything more extreme**, if the null hypothesis is true. By "anything more extreme", we mean any result that would alert us even more (than the result we observed) to the possibility that our research hypothesis, and not the null hypothesis, is true. In our example, the research hypothesis is that the mean weight is less than 200 lb.; therefore a sample mean less than 200 lb. (which would result in a negative value of t) could suggest that the research hypothesis is true, and any value of t less than (i.e., more negative than) −1.9 would alert us even more to the possibility that the null hypothesis is not true. A t-value of + 2.5, on the other hand, would certainly not suggest that the research hypothesis is true.

## ONE-SIDED VERSUS TWO-SIDED TESTS

Now suppose, in the above example, we had wished to determine whether the mean weight **is different from** 200 lb., rather than **is less than** 200 lb. Expressed in this way, our research hypothesis is that there is a difference, but in an unspecified direction. We believe that the program will affect weight, but are unwilling to state ahead of time whether the final weight will be more or less than 200 lb. Any extreme deviation from 200 lb., whether positive or negative, would suggest that the null hypothesis is not true. Had this been the case, not only would a t-value less than −1.9 be more extreme, but so also would any t-value **greater than** +1.9. Thus, because of the symmetry of the t distribution, the value of p would be **double** 4.5%, that is,

9%: we add up the probability to the left of −1.9 and the probability to the right of +1.9 (i.e., the probabilities in both tails of the distribution).

Thus the significance level depends on what we had in mind before we actually sampled the population. If we knew beforehand that the weight-reduction program could not lead to a weight above 200 lb., our question would be whether or not the mean weight is *less* than 200 lb. We would perform what is known as a **one-sided**, or **one-tail**, test, using only the left-hand tail of the t distribution; and we would report the resulting t = −1.9 as being significant at the 5% level. If, on the other hand, we had no idea originally whether the program would lead to a mean weight above or below 200 lb., the question of interest would be whether or not the true mean is *different from* 200 lb. We would then perform a **two-sided** or **two-tail** test, using the probabilities in both the left- and right-hand tails of the t distribution; and for our example, a t-value of −1.9 would then not be significant at the 5% level, although it would be significant at the 10% level (p $\doteq$ 0.09).

There is a close connection between a two-sided test and a confidence interval. Let us calculate the 95% and 90% confidence intervals for the mean weight of men on the weight-reduction program. We have

$$n = 10, \bar{y} = 184, s_{\bar{y}} = 8.38.$$

From Table 1, Appendix 2, we see that for 9 degrees of freedom $t_{97.5}$ = 2.262 and $t_{95}$ = 1.833. We therefore have

95% confidence interval: 184 ± (2.262)8.38, or 165.04 to 202.96.
90% confidence interval: 184 ± (1.833)8.38, or 168.64 to 199.36.

The 95% interval includes the value 200, whereas the 90% interval does not. In general, a sample estimate (184 in this example) will be significantly different from a hypothesized value (200) if and only if the corresponding confidence interval for the parameter does **not** include that value. A 95% confidence interval corresponds to a two-sided test at the 5% significance level: the interval contains 200, and the test is not significant at the 5% level. A 90% confidence interval corresponds to a test at the 10% significance level: the interval does not include 200, and the test is significant at the 10% level. In general a $100(1-\alpha)\%$ confidence interval corresponds to a two-sided test at the $100\alpha\%$ significance level.

## TESTING A HYPOTHESIS ABOUT A PROPORTION

Suppose an investigator disputes a claim that, using a new surgical procedure for a risky operation, the proportion of successes is at least 0.6. The investigator is interested in showing that the proportion is in fact less than 0.6. The three steps to test this research hypothesis are then as follows:

1. The null hypothesis is taken to be that the proportion of successes is 0.6 (i.e., $\pi = 0.6$); we shall see whether the data are consistent with this null hypothesis, or whether we should reject it in favor of $\pi < 0.6$.
2. Suppose Y, the number of successes, is binomially distributed. Then we can use Y as our test criterion: once the sample size is determined, its distribution is known if in fact $\pi = 0.6$ and Y is binomially distributed.
3. We select a sample of operations in which this new procedure is used, say n = 10 operations, and find, let us suppose, y = 3 successes. From the binomial distribution with n = 10 and $\pi = 0.6$, the probability of each possible number of successes is as follows

| Number of Successes | Probability |
|:---:|:---:|
| 0 | 0.0001 |
| 1 | 0.0016 |
| 2 | 0.0106 |
| 3 | 0.0425 |
| 4 | 0.1114 |
| 5 | 0.2007 |
| 6 | 0.2508 |
| 7 | 0.2150 |
| 8 | 0.1209 |
| 9 | 0.0404 |
| 10 | 0.0060 |
| Total | 1.0000 |

To determine the p-value, we sum the probabilities of all outcomes as extreme or more extreme than the one observed. The "extreme outcomes" are those that suggest the research hypothesis is true and alert us, even more than the sample itself, to the possibility that the null hypothesis is false. If $\pi = 0.6$, we expect, on an average, 6 successes in 10 operations. A series of 10 operations with fewer successes would suggest that $\pi < 0.6$, and hence that the research hypothesis is true and the null hypothesis is false. Thus, 0, 1, 2, or 3 successes would be as extreme or more extreme than the 3 observed. We sum the probabilities of these four outcomes to obtain $0.0001 + 0.0016 + 0.0106 + 0.0425 = 0.0548$ (i.e., p = 0.0548). We find it difficult to believe that we would be so unlucky as to obtain an outcome as rare as this if $\pi$ is at least 0.6, as claimed. We believe, rather, that $\pi < 0.6$, since this would give rise to larger probabilities of observing 0, 1, 2, or 3 successes in 10 operations. We are therefore inclined to reject the null hypothesis and conclude that the probability of success using the new procedure is less than 0.6. Specifically, we can say that the observed proportion of successes, 3 out of 10, or 0.3, is significantly less than 0.6 at the 6% level.

Let us suppose, for illustrative purposes, that the sample of 10 operations had resulted in $y = 8$ successes. Such an outcome would be consistent with the null hypothesis. All y values less than 8 would be closer to the research hypothesis than the null hypothesis, and so the p-value for such an outcome would be

$$\begin{aligned} P(0) + P(1) + \ldots + P(8) &= 1 - P(9) - P(10) \\ &= 1 - 0.0404 - 0.0060 \\ &= 0.9536 \ . \end{aligned}$$

In this instance it is obvious that we should retain our hypothesis (i.e., the data are consistent with the hypothesis that the probability of a success using the new procedure is at least 0.6). But note carefully that "being consistent with" a hypothesis is not the same as "is strong evidence for" a hypothesis. We would be much more convinced that the hypothesis is true if there had been 800 successes out of 1,000 operations. "Retaining" or "accepting" the null hypothesis merely means that we have insufficient evidence to reject it—not that it is true.

If the number of operations had been large, much effort would be needed to calculate, from the formula for the binomial distribution, the probability of each possible more extreme outcome. Suppose, for example, there had been n = 100 operations and the number of successes was y = 30, so that the proportion of successes in the sample is still 0.3, as before. In this case it would be necessary to calculate the probabilities of 0, 1, 2, . . . right on up to 30 successes, in order to obtain the exact p-value. But in such a case we can take advantage of the fact that n is large. When n is large we know that the average number of successes per operation, Y/n (i.e., the proportion of successes) is approximately normally distributed. The three steps in testing the hypothesis are then as follows:

1. The null hypothesis is $\pi = 0.6$, as before.
2. Since both $n\pi$ and $n(1 - \pi)$ are greater than 5 under the null hypothesis (they are 60 and 40, respectively), we assume that Y/n is normally distributed with mean 0.6 and standard deviation $\sqrt{0.6(1-0.6)/n} = \sqrt{0.24/100} = 0.049$. Thus, under the null hypothesis, the standardized variable

$$Z = \frac{Y/n - 0.6}{0.049}$$

   approximately follows a standard normal distribution and can be used as the test criterion.
3. We observe y/n = 0.3 and hence

$$z = \frac{0.3 - 0.6}{0.049} = -6.12$$

and any value of z less than this is more extreme (i.e., even less consistent with the null hypothesis). Consulting the probabilities for the standard normal distribution (last line of Table 1, Appendix 2), we find

$$P(Z < -2.576) = 0.005,$$

and $P(Z < -6.12)$ must be even smaller than this. We are thus led to reject the null hypothesis at an even smaller significance level.

We can see the improvement in the normal approximation of the binomial distribution with increasing sample sizes from the following probabilities, calculated on the assumption that $\pi = 0.6$

| | $P(Y/n \leq 0.3)$ | |
|:---:|:---:|:---:|
| Sample size | Binomial | Normal |
| 10 | 0.0548 | 0.0262 |
| 20 | 0.0065 | 0.0031 |
| 30 | 0.0002 | 0.0004 |

If we restrict our attention to the first two decimal places, the difference in p-values is about 0.03 for a sample of size 10, but less than 0.01 for a sample of size 20 or larger.

The binomial distribution, or the normal approximation in the case of a large sample, can be used in a similar manner to test hypotheses about any percentile of a population distribution. As an example, suppose we wish to test the hypothesis that the median (i.e., the 50th percentile) of a population distribution is equal to a particular hypothetical value. A random sample of n observations from the distribution can be classified into two groups: those above the hypothetical median and those below the hypothetical median. We then simply test the null hypothesis that the proportion above the median (or equivalently, the proportion below the median) is equal to 0.5, (i.e., $\pi = 0.5$). This is simply a special case of testing a hypothesis about a proportion in a population. If we ask whether the population median is **smaller** than the hypothesized value, we perform a one-sided test similar to the one performed above. If we ask whether it is **larger**, we similarly perform a one-sided test, but the p-value is obtained by summing the probabilities in the other tail. If, finally, we ask whether the median is **different** from the hypothesized value, a **two-sided** test is performed, summing the probabilities of the extreme outcomes in both tails.

## TESTING THE EQUALITY OF TWO VARIANCES

Often, we wish to compare two samples. We may ask, for example, whether the distribution of serum cholesterol levels is the same for males and females in a set of patients. First we could ask whether the distribution in each population is normal, and there are various tests for this. If we find the assumption of normality reasonable, we might then assume normality and ask whether the variance is the same in both populations from which the samples come. Let the two sample variances be $s_1^2$ and $s_2^2$. Then an appropriate criterion to test the null hypothesis that the two population variances are equal is the ratio $s_1^2/s_2^2$. Provided the distribution in each population is normal, under the null hypothesis this statistic comes from an F distribution, named in honor of Sir Ronald Fisher. The F distribution is a two-parameter distribution, the two parameters being the number of degrees of freedom in the numerator $(s_1^2)$ and the number of degrees of freedom in the denominator $(s_2^2)$. If the sample sizes of the two groups are $n_1$ and $n_2$, then the numbers of degrees of freedom are respectively $n_1 - 1$ and $n_2 - 1$. A few percentile points of the F distribution are given in Table 2 of Appendix II. The number of degrees of freedom along the **top** of the table corresponds to that in the **top** of the F-ratio $(n_1 - 1)$, whereas that along the side of the table corresponds to that in the bottom of the F-ratio. (All tables of the F distribution follow this convention.) The table is appropriate for testing the null hypothesis $\sigma_1^2 = \sigma_2^2$ against the alternative $\sigma_1^2 > \sigma_2^2$: large values of F are significant. This is a one-sided test. If we wish to perform a two-sided test, we put the larger of the two sample variances, $s_1^2$ or $s_2^2$, on top and **double** the tail probability indicated by the table.

A numerical example will illustrate the procedure. Suppose we have a sample of $n_1 = 10$ men and $n_2 = 25$ women, with sample variances $s_1^2 = 30.3$ and $s_2^2 = 69.7$, respectively, for a trait of interest. We wish to test the null hypothesis that the two population variances are equal (i.e., $\sigma_1^2 = \sigma_2^2$). We have no prior knowledge to suggest which might be larger, and so we wish to perform a two-sided test. We therefore put the larger sample variance on top to calculate the ratio

$$\frac{69.7}{30.3} = 2.30$$

There are $25 - 1 = 24$ degrees of freedom in the top of this ratio and $10 - 1 = 9$ degrees of freedom in the bottom. Looking at the columns headed 24, and the rows labelled 9, in the table of the F distribution (Table 2, Appendix 2), we find the following percentiles

| %: | 90 | 95 | 97.5 | 99 | 99.5 |
|---|---|---|---|---|---|
| F-value: | 2.28 | 2.90 | 3.61 | 4.73 | 5.73 |

The observed ratio, 2.30, lies between the 90th and 95th percentiles, corresponding to tail probabilities of 0.1 and 0.05. Because we wish to perform a two-sided test, we double these probabilities to obtain the p-value. The result would thus be quoted as $0.1 < p < 0.2$, or as $p \doteq 0.2$ (since 2.30 is close to 2.28).

In this instance we might decide it is reasonable to assume that the two variances are equal. As we learned in Chapter 6, the common, or "pooled," variance is then estimated as

$$s_p^2 = \frac{(n_1 - 1)s_1^2 + (n_2 - 1)s_2^2}{n_1 + n_2 - 2},$$

which in this case is

$$\frac{(9)30.3 + (24)69.7}{9 + 24} = 59.0 .$$

This estimate is unbiased.

Note once again that the null hypothesis is not simply that the two variances are equal, although the F-test is often described as a test for the equality of two variances. For the test criterion to follow an F distribution, each sample must also be made up of normally distributed random variables. In other words, the null hypothesis is that the two samples are made up of independent observations from two normally distributed populations with the same variance. The distribution of the F-statistic is known to be especially sensitive to non-normality, so a significant result could be due to non-normality and have nothing to do with whether or not the population variances are equal.

## TESTING THE EQUALITY OF TWO MEANS

Suppose now we can assume that the random variables of interest are normally distributed, with the same variance in the two populations. Then we can use a **two-sample t-test** to test whether the means of the random variable are significantly different in the two populations. Let $\overline{Y}_1$ and $\overline{Y}_2$ be the two sample means. Then, under the null hypothesis that the two population means are the same, $\overline{Y}_1 - \overline{Y}_2$ will be normally distributed with mean zero. Furthermore, provided the observations in the two groups are independent (taking separate random samples from the two populations will ensure this), the variance of $\overline{Y}_1 - \overline{Y}_2$ will be $\sigma_{\overline{Y}_1}^2 + \sigma_{\overline{Y}_2}^2$ (i.e., $\sigma^2/n_1 + \sigma^2/n_2$) where $\sigma^2$ is the common variance. Thus

$$\frac{\overline{Y}_1 - \overline{Y}_2}{\sqrt{\sigma_{\overline{Y}_1}^2 + \sigma_{\overline{Y}_2}^2}} = \frac{\overline{Y}_1 - \overline{Y}_2}{\sigma\sqrt{\dfrac{1}{n_1} + \dfrac{1}{n_2}}}$$

will follow a standard normal distribution, and analogously,

$$\frac{\overline{Y}_1 - \overline{Y}_2}{S_p\sqrt{\dfrac{1}{n_1} + \dfrac{1}{n_2}}}$$

will follow a t distribution with $n_1 + n_2 - 2$ degrees of freedom; in this formula we have replaced $\sigma$ by $S_p$, the square root of the pooled variance estimator. Thus we calculate t, and compare it with percentiles of the t distribution with $n_1 + n_2 - 2$ degrees of freedom. As before, if $n_1 + n_2 - 2$ is greater than 30, the percentiles are virtually the same as for the standard normal distribution.

Suppose, for our example of $n_1 = 10$ men and $n_2 = 25$ women, we found the sample means $y_1 = 101.0$ and $y_2 = 97.5$. We have already seen that $s_p^2 = 59.0$, and so $s_p = \sqrt{59.0} = 7.68$. To test whether the means are significantly different, we calculate

$$\frac{y_1 - y_2}{s_p\sqrt{\dfrac{1}{n_1} + \dfrac{1}{n_2}}} = \frac{101.0 - 97.5}{7.68\sqrt{\dfrac{1}{10} + \dfrac{1}{25}}} = 2.04.$$

There are $9 + 24 = 33$ degrees of freedom, so 2.04 is approximately the 97.5th percentile of the t distribution. Thus $p \doteq 0.025$ for a one-sided test and $p \doteq 0.05$ for a two-sided test.

Note carefully the assumption that the two samples are independent. Often this assumption is purposely violated in designing an experiment to compare two groups. Cholesterol levels, for example, change with age; so if our sample of men were very different in age from our sample of women, we would not know whether any difference that we found was due to gender or to age (i.e., these two effects would be confounded). To obviate this, we could take a sample of men and women who are individually matched for age. We would still have two samples, n men and n women, but they would no longer be independent. We would expect the pairs of cholesterol levels to be correlated, in the sense that the cholesterol levels of a man and woman who are the same age will tend to be more alike than those of a man and woman who are different ages (the term "correlation" will be defined more precisely in Chapter 9). In this case an appropriate test for a mean difference between the two populations would be the **paired t-test.** We would pair the men and women and find the difference in cholesterol level for each pair (taking care always to subtract the cholesterol level of a person of the same sex—so some of the differences may be negative). Let the difference be d. Then we have n d's, and, if the null hypothesis of no mean difference between male and female is true, the d's are expected to have mean zero. Thus we would calculate

$$t = \frac{\overline{d}}{s_{\overline{D}}} = \frac{\overline{d}}{s_D/\sqrt{n}} \, ,$$

where $\overline{d}$ is the mean of the n d's and $s_D$ their estimated standard deviation, and compare this with percentiles of the t distribution with $n - 1$ degrees of freedom. This test assumes that the differences (the d's) are normally distributed.

## TESTING THE EQUALITY OF TWO MEDIANS

The median of a normal distribution is the same as its mean. It follows that if our sample data come from normal distributions, testing for the equality of two medians is the same as testing for the equality of two means. If our samples do not come from normal distributions, however, we should not use the t distribution as indicated above to test for the equality of two means in small samples. Furthermore, if the population distributions are at all skewed, the medians are better parameters of central tendency. We should probably be more interested in testing the equality of the two medians than in testing the equality of the two population means. In this section we shall outline methods of doing this without making distributional assumptions, such as normality. For this reason the methods we shall describe are sometimes called **distribution-free** methods. We shall indicate statistics that can be used as criteria for the tests and note that for large samples they are approximately normally distributed, regardless of the distributions of the underlying populations. It is beyond the scope of this book to discuss the distribution of all these statistics in small samples, but you should be aware that appropriate tables are available for such situations.

First, suppose we have two independent samples: $n_1$ observations from one population and $n_2$ observations from a second population. The **rank sum test** is the appropriate test in this situation, provided we can assume that the distributions in the two populations, while perhaps having different medians, have the same (arbitrary) shape. The observations in the two samples are first considered as a single set of $n_1 + n_2$ numbers, and arranged in order from the smallest to largest. Each observation is assigned a rank: 1 for the smallest observation, 2 for the next smallest, and so on, until $n_1 + n_2$ is assigned to the largest observation. The ranks of the observations in the smaller of the two samples are then summed, and this is the statistic, which we denote T, whose distribution is known under the null hypothesis. Percentile points of the distribution of T have been tabulated, but for large samples we can assume that T is approximately normally distributed.

As an example, suppose we wish to compare the median serum cholesterol levels in mg/100 ml for two groups of students, based on the following samples:

Sample 1, $n_1 = 6$: 58 , 92 , 47 , 126 , 53, 85
Sample 2, $n_2 = 7$: 87 , 199 , 124 , 83 , 115, 68 , 156

The combined set of numbers and their corresponding ranks are:

| 47 | 53 | 58 | 68 | 83 | 85 | 87 | 92 | 115 | 124 | 126 | 156 | 199 |
|----|----|----|----|----|----|----|----|-----|-----|-----|-----|-----|
| 1  | 2  | 3  | 4  | 5  | 6  | 7  | 8  | 9   | 10  | 11  | 12  | 13  |

The underlined ranks correspond to the smaller sample, $1 + 2 + 3 + 6 + 8 + 11 = 31$. Although the samples are not really large enough to justify using the large sample normal approximation, we shall nevertheless use these data to illustrate the method. We standardize T by subtracting its mean and dividing by its standard deviation, these being derived under the null hypothesis that the two medians are equal. The result is then compared with percentiles of the normal distribution. If $n_1 \leq n_2$ (i.e., $n_1$ is the size of the smaller sample) it is shown in Appendix I that the mean value of T is $n_1(n_1 + n_2 + 1)/2$, which in this case is $6(6 + 7 + 1)/2 = 42$. Also, it can be shown that the standard deviation of T is $\sqrt{n_1 n_2 (n_1 + n_2 + 1)/12}$, which in our example is $\sqrt{(6)(7)(14)/12} = 7$. Thus we calculate the standardized criterion

$$z = (31 - 42)/7 = 1.57.$$

Looking this up in the table (last line of Table 1, Appendix 2), we find it lies between the 90th and 95th percentiles. In fact, tables that give percentiles of the exact distribution of T also indicate that $0.1 < p < 0.2$ for a two-sided test, so in this instance the normal approximation does not mislead us.

Let us now suppose the samples were taken in such a way that the data are paired, with each pair consisting of one observation from each population. The study units might be paired, for example, in a randomized-blocks experimental design in which each block consists of only two subjects (of the same age and sex) randomly assigned to one or the other of two treatments. Paired observations could also arise in situations in which the same subject is measured before and after treatment.

Since the study units are paired, the difference in the observation of interest can be computed for each pair, taking care always to calculate the difference in the same direction. These differences can then be analyzed by the **signed rank sum test** as follows: First we rank the differences from smallest to largest, without regard to the sign of the difference. Then we sum the ranks of the positive and negative differences separately, and the smaller of these two numbers is entered into an appropriate table to determine the p-value. For large samples we can again use a normal approximation, using the fact that under the null hypothesis the mean and the standard deviation of the sum depend only on the number of pairs, n.

As an example of this test, let us suppose that eight identical twin pairs were studied to investigate the effect of a triglyceride-lowering drug. A member of each pair was randomly assigned to either the active drug or a placebo, with the other member of the pair receiving the other treatment. The resulting data are as follows (triglyceride values are in mg/100 ml)

| Twin pair | 1 | 2 | 3 | 4 | 5 | 6 | 7 | 8 |
|---|---|---|---|---|---|---|---|---|
| Placebo twin | 71 | 65 | 126 | 111 | 249 | 198 | 57 | 97 |
| Active drug twin | 69 | 52 | 129 | 75 | 226 | 181 | 46 | 93 |
| Difference | 2 | 13 | −3 | 36 | 23 | 17 | 11 | 4 |
| Rank (ignoring sign) | 1 | 5 | 2 | 6 | 8 | 7 | 4 | 3 |

The sum of the ranks of the positive differences is $1 + 5 + 6 + 8 + 7 + 4 + 3 = 34$, and that of the negative differences (there is only one) is 2. If we were to look up 2 in the appropriate table for n = 8, we would find, for a two-tail test, $0.02 < p < 0.05$. Hence, we would reject the hypothesis of equal medians at $p = 0.05$, and conclude that the active drug causes a (statistically) significant reduction in the median triglyceride level.

The large sample approximation will now be computed for this example, to illustrate the method. Under the null hypothesis the mean of the sum is $n(n+1)/4$ and the standard deviation is $\sqrt{n(n+1)(2n+1)/24}$. Thus, when n = 8, the mean is $8(9)/4 = 18$ and the standard deviation is $\sqrt{8(9)(17)/24} = \sqrt{51} = 7.14$.

We therefore calculate

$$z = \frac{2 - 18}{7.14} = -2.24 ,$$

which, from the normal table (last line of Table 1, Appendix 2), lies between the 1st and 2.5th percentiles. For a two-sided test this corresponds to $0.02 < p < 0.05$. Thus, even for as small a sample as this, we once again find that the normal approximation is adequate.

Let us now briefly consider another way of testing the same hypothesis. If the medians are equal in the two populations, then on an average, the number of positive differences in the sample will be the same as the number of negative differences. In other words, the mean proportion of positive differences would be 0.5 in the population (of all possible differences). Thus, we can test the null hypothesis that the proportion of positive differences, $\pi$, is 0.5. This is called the **sign test**. For a sample size n = 8 as in the above data, we have the following binomial distribution under the null hypothesis ($\pi = 0.5$):

| Number of minus signs | Probability |
|---|---|
| 0 | 0.0039 |
| 1 | 0.0313 |
| 2 | 0.1093 |
| 3 | 0.2188 |
| 4 | 0.2734 |
| 5 | 0.2188 |
| 6 | 0.1093 |
| 7 | 0.0313 |
| 8 | 0.0039 |

Thus, the probability of observing a result as extreme or more extreme than a single minus sign under the null hypothesis is

$$P(0) + P(1) + P(7) + P(8) = 0.0039 + 0.0313$$
$$+ 0.0313 + 0.0039 = 0.0704.$$

(We sum the probabilities in both tails, for a two-sided test). This result, unlike the previous one based on the same data, is no longer significant at the 5% significance level. Which result is correct, this one or the previous one? Can both be correct? To understand the difference we need to learn about how we judge different hypothesis-testing procedures.

## VALIDITY AND POWER

Sometimes we have to make a definite decision one way or another about a particular hypothesis. Although in science we never accept a hypothesis outright, but rather continually modify our ideas and laws as new knowledge is obtained, in clinical practice we cannot afford this luxury. To understand the concepts of validity and power it will be helpful if we consider the case in which a decision must be made, one way or the other, with the result that some wrong decisions will inevitably be made. Clearly we wish to act in such a way that the probability of making a wrong decision is minimized.

Let us suppose we perform a statistical test and on the basis of that test decide either to accept or to reject the null hypothesis, for which from now on we shall use the abbreviation $H_0$. Since in the "true state of nature" $H_0$ must be actually true or false, we have just four possibilities, which we can depict as the entries in a $2 \times 2$ table as follows

|  |  | True state of nature | |
| --- | --- | --- | --- |
|  |  | $H_0$ is true | $H_0$ is false |
| Decision made | Accept $H_0$ | O.K. | Type II error |
|  | Reject $H_0$ | Type I error | O.K. |

In the case of two of the possibilities, denoted by the entries "O.K.," the decision is correct, and hence no error is made. In the case of the other two possibilities, a wrong decision, and hence an error, is made. The error may be one of two types:

Type I.  Rejection of the null hypothesis when in fact it is true. The probability of this happening is often denoted $\alpha$ [i.e., $\alpha = P(\text{reject } H_0 \mid H_0 \text{ is true})$].

Type II. Acceptance of the null hypothesis when in fact it is false. The probability of this happening is often denoted $\beta$ [i.e., $\beta = P(\text{accept } H_0 \mid H_0 \text{ is false})$].

When using a significance test to make a decision of this sort, we

1. decide **beforehand** on some particular value of $\alpha$ (e.g., 0.05), the **significance level.**
2. experiment, calculate the test statistic, and determine whether or not it is significant at the $100\alpha\%$ (e.g., 5%) level.
3. accept the null hypothesis if it is **not** significant; reject the null hypothesis if it **is** significant.

In this procedure our significance level, $\alpha$, is the probability of making a type I error, and we control it so that it is kept reasonably small. Suppose, for example, we decide to fix $\alpha$ at the value 0.05. It is our intention to accept the null hypothesis if the result is not significant at the 5% level and to reject it if it is. Then, provided our test does in fact reject $H_0$ in 5% of the situations in which $H_0$ is true, it is a **valid** test at the 5% level. A valid test is one that rejects $H_0$ in a proportion $\alpha$ of the situations in which $H_0$ is true, where $\alpha$ is the stated significance level.

Suppose we have a sample of paired data from two populations that are normally distributed with the same variance. In order to test whether the two population medians are equal, we could use: (1) the paired t-test; (2) the signed rank sum test; or (3) the sign test. The fact that we have normal distributions does not in any way invalidate the signed rank sum and the sign tests. Provided we use the appropriate percentiles of our test criteria (e.g., the 5th percentile or the 95th percentile for a one-sided test) to determine whether or not to accept the null hypothesis, we shall find that, **when it is true**, we reject $H_0$ with 5% probability. This will be true of all three tests; they are all **valid** tests in this situation.

Although they are all valid, the three tests nevertheless differ in the value of $\beta$, or probability of Type II error. In other words, they differ in the probability of accepting the null hypothesis **when it is false** (i.e., when the medians are in fact different). In this situation we are most likely to reject the null hypothesis when using the t-test, less likely to do so when using the signed rank sum test, and least likely to do so when using the sign test. We say that the t-test is more **powerful** than the signed rank sum test, and the signed rank sum test is more powerful than the sign test. Power is defined as $1 - \beta$: it is the probability of rejecting the null hypothesis when in fact it is false. Note that if we identify the null hypothesis with absence of a disease, there is an analogy between the power of a statistical test and the sensitivity of a diagnostic test (defined in Chapter 3).

Now suppose we do not have normal distributions, but we can assume that the shape of the distribution is the same in both populations. If this is the case, the paired t-test may no longer be valid, and then the fact that it might be more powerful is irrelevant. Now in large samples, the t-test is fairly robust against non-normality (i.e., the test is approximately valid even when we do not have underlying normal distributions). But this is not necessarily the case in small samples. We should not use the t-test for small samples if there is any serious doubt about the underlying population distributions being about normal. Note that if the samples are large enough for the t-test to be robust, then we do not need to refer the test statistic to the t distribution. We saw in the last chapter that the t distribution with more than 30 degrees of freedom has percentiles that are about the same as those of the standard normal distribution.

If we cannot assume that the shape of the distribution is about the same in both populations, then both the paired t-test and the signed rank-sum test may be invalid, and we should use the sign test even though it is the least powerful. This illustrates a general principle of all statistical tests: The more we can assume, the more powerful our test can be. This same principal is at work in the distinction between one-sided and two-sided tests. If we are prepared to assume, prior to any experimentation, that the median of population 1 cannot be smaller than that of population 2, we can perform a one-sided test. Then, to attain a p-value less than a pre-specified amount, our test criterion need not be as extreme as would be necessary for the analogous two-sided test. Thus a one-sided test is always more powerful than the corresponding two-sided test.

We often do not control $\beta$, the probability of making an error if $H_0$ is false, mainly because there are many ways in which $H_0$ can be false. It makes sense, however, to have some idea of the magnitude of this error before going to the expense of conducting an experiment. This is done by calculating $1 - \beta$, the **power** of the test, and plotting it against the "true state of nature." Just as the sensitivity of a diagnostic test to detect a disease usually increases with the severity of the disease, so the power of a statistical test usually increases with departure from $H_0$. For example, for the two-sided t-test of the null hypothesis $\mu_1 = \mu_2$, we can plot power against $\mu_1 - \mu_2$, as in Figure 7.2. Note that the "power curve" is symmetrical about $\mu_1 - \mu_2 = 0$ (i.e., about $\mu_1 = \mu_2$), since we are considering a two-sided test. Note also that the probability of rejecting $H_0$ is a minimum when $H_0$ is true (i.e., when $\mu_1 - \mu_2 = 0$), and that at this point it is equal to $\alpha$, the significance level. The power increases as the absolute difference between $\mu_1$ and $\mu_2$ increases (i.e., as the departure from $H_0$ increases).

As you might expect, power also depends on the sample size. We can always make the probability of rejecting $H_0$ small by studying a small sample. Hence, not finding a significant difference, or "accepting $H_0$," must never be equated with the belief that $H_0$ is true: it

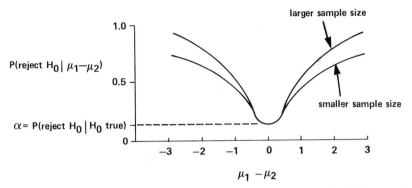

**FIGURE 7.2** Examples of the power of the two-sided t-test for the difference between two means, $\mu_1$ and $\mu_2$, plotted against $\mu_1 - \mu_2$.

merely indicates that there is insufficient evidence to reject $H_0$ (which may be due to the fact that $H_0$ is true, **or** may be due to a small sample size). It is possible to determine from the power curve how large the difference $\mu_1 - \mu_2$ must be in order for there to be a good chance of rejecting $H_0$ (i.e., of observing a difference that is statistically significant). Also, we could decide on a magnitude for the real difference that we should like to detect, and then plot against sample size the power of the test to detect a difference of that magnitude. This is often done before conducting a study, in order to choose an appropriate sample size. Power also depends on the variability of our measurements, however; the more variable they are, the less the power. For this reason power is often expressed as a function of the standardized difference $(\mu_1 - \mu_2)/\sigma$, where it is assumed that the two populations have the same standard deviation. For example, a small difference is often considered to be less than $0.2\sigma$, a medium difference between $0.2\sigma$ and $0.8\sigma$, and a large difference one that is larger than $0.8\sigma$.

In summary, there are six ways to increase power when testing a hypothesis statistically:

1. Use a larger significance level. This is often less desirable since it implies a larger probability of Type I error.
2. Use a larger sample size. This is more expensive.
3. Consider only larger deviations from $H_0$. This may be less desirable; but note there is no point in considering differences that are too small to be biologically or medically significant.

4. Reduce variability, either by making more precise measurements or by choosing more homogeneous study units.

5. Make as many valid assumptions as possible (e.g., a one-sided test is more powerful than a two-sided test).

6. Use the most powerful test that the appropriate assumptions will allow. The most powerful test may sometimes be more expensive to compute, but it is usually cheapest in the long run.

It is also possible to increase power by using an invalid test, but this is never legitimate!

Finally, remember that if a statistical test shows that a sample difference is not significant, this does not prove that a population difference does not exist, or even that any real difference is probably small. Only the **power** of the test tells us anything about the probability of rejecting any hypothesis other than the null hypothesis. Thus, we are fairly safe in assuming that we have learned something from our data when we reject the null hypothesis. When we accept the null hypothesis we are not so well-informed, and only a careful analysis of the power will determine whether we have learned anything at all.

## SUMMARY

1. The three steps in a significance test are: (1) determine a specific null hypothesis to be tested; (2) determine an appropriate test criterion—a statistic whose sampling distribution is known under the null hypothesis; and (3) calculate the test criterion from the sample data and determine the corresponding significance level.

2. The p-value is the probability of observing what we actually did observe, or anything more extreme, if the null hypothesis is true. The result is significant at any level larger than or equal to p, but not significant at any level less than p. This is **statistical** significance, as opposed to **biological** or **medical** significance.

3. In a one-sided (one-tail) test, results that are more extreme in one direction only are included in the evaluation of p. In a two-sided test, results that are more extreme in both directions are included. Thus, to attain a specified significance level, the test statistic need be less "atypical" for a one-sided test than for a two-sided test.

4. Hypotheses about a proportion or percentile can be tested using the binomial distribution. For large samples, an observed proportion

is about normally distributed with mean $\pi$ and standard deviation $\sqrt{\pi(1 - \pi)/n}$.

5. To test the equality of two variances we use the F-statistic: the ratio of the two sample variances; the number of degrees of freedom in the top of the ratio corresponds to that along the top of the F-table, the number in the bottom corresponds to that along the side of the table. The larger sample variance is put at the top of the ratio; for a two-sided test, the tail probability indicated by the table is doubled. The F-statistic is sensitive to non-normality.

6. If we have normal distributions, the t distribution can be used to test for the equality of two means. If we have two independent samples of sizes $n_1$ and $n_2$ from populations with the same variance, we use the two-sample t-test after estimating a pooled variance with $n_1 + n_2 - 2$ degrees of freedom. If we have a sample of n correlated pairs of observations, we use the n differences as a basis for the paired t-test, with $n - 1$ degrees of freedom.

7. If we have two populations with similarly shaped distributions, the rank sum test can be used to test the equality of the two medians when we have two independent samples, and the signed rank sum test when we have paired data. The sign test can also be used for paired data without making any assumption about the underlying distributions. All three tests are based on statistics that, when standardized, are about normally distributed in large samples.

8. A valid test is one for which the stated probability of the Type I error (p-value) is correct: when the null hypothesis is true it leads to rejection of the null hypothesis with probability p. A powerful test is one for which the probability of Type II error, (i.e., the probability of accepting the null hypothesis when it is false), is low.

9. The more assumptions that can be made, the more powerful a test can be. A one-sided test is more powerful than a two-sided test. The power of a statistical test can also be increased by using a larger significance level, a larger sample size, or by deciding to try to detect a larger difference; it is decreased by greater variability, whether due to measurement error or heterogeneity of study units.

## FURTHER READING

Altman D.G. (1980) Statistics and Ethics in Medical Research: III How Large a Sample? British Medical Journal 281:1336–1338. (This article contains a nomogram, for a two-sample t-test with equal numbers in each sample, relating power, total study size, the standardized mean difference, and the significance level. Given any three of these quantities, the fourth can be read off the nomogram.)

Blackwelder W.C. (1982) "Proving the Null Hypothesis" in Clinical Trials. Controlled Clinical Trials 3:345–353. (This article shows how to set up the statistical null hypothesis in a situation in which the research hypothesis of interest is that two different therapies are equivalent.)

## PROBLEMS

1. Significance testing and significance levels are important in the development of science because
   A. they allow one to prove a hypothesis is false
   B. they provide the most powerful method of testing hypotheses
   C. they allow one to quantify one's believe in a particular hypothesis other than the null hypothesis
   D. they allow one to quantify the likelihood of a sample result if the null hypothesis is false
   E. they allow one to quantify the likelihood of a sample result if the null hypothesis is true

2. A one-sided test to determine the significance level is particularly relevant for situations in which
   A. we have paired observations
   B. we know *a priori* the direction of any true difference
   C. only one sample is involved
   D. we have normally distributed random variables
   E. we are comparing just two samples

3. If a one-sided test indicates that the null hypothesis can be rejected at the 5% level, then
   A. the one-sided test is necessarily significant at the 1% level
   B. a two-sided test on the same set of data is necessarily significant at the 5% level
   C. a two-sided test on the same set of data cannot be significant at the 5% level
   D. a two-sided test on the same set of data is necessarily significant at the 10% level
   E. the one-sided test cannot be significant at the 1% level

4. A researcher conducts a clinical trial to study the effectiveness of a new treatment in lowering blood pressure and concludes that "the lowering of mean blood pressure in the treatment group was significantly greater than that in the group on placebo (p < 0.01)." This means that

   A. if the treatment has no effect, the probability of the treatment group having a lowering in mean blood pressure as great as or greater than that observed is exactly 1%

   B. if the treatment has no effect, the probability of the treatment group having a lowering in mean blood pressure as great as or greater than that observed is no greater than 1%

   C. there is exactly a 99% probability that the treatment lowers blood pressure

   D. there is at least a 99% probability that the treatment lowers blood pressure

   E. none of the above.

5. A surgeon claims that at least 3/4 of his operations for gastric resection are successes. He consults a statistician and together they decide to conduct an experiment involving 10 patients. Assuming the binomial distribution is appropriate, the following probabilities are of interest:

| Number of Successes | Probability of Success with $p = 3/4$ |
|---|---|
| 0 | 0.0000 |
| 1 | 0.0000 |
| 2 | 0.0004 |
| 3 | 0.0031 |
| 4 | 0.0162 |
| 5 | 0.0582 |
| 6 | 0.1460 |
| 7 | 0.2503 |
| 8 | 0.2816 |
| 9 | 0.1877 |
| 10 | 0.0563 |

Suppose 4 of the 10 operations are successes. Which of the following conclusions is best?

   A. The claim should be doubted, since the probability of observing 4 or fewer successes with $p = 3/4$ is 0.0197.

B. The claim should not be doubted, since the probability of observing 4 or more successes is 0.965.
C. The claim should be doubted only if 10 successes are observed.
D. The claim should be doubted only if no successes are observed.
E. None of the above.

6. "The difference is significant at the 1% level" implies
(1) there is a 99% probability that there is a real difference
(2) there is at most a 1% probability of something as or more extreme than the observed result occurring if, in fact, the difference is zero
(3) the difference is significant at the 0.1% level
(4) the difference is significant at the 5% level
  A. 1, 2, and 3
  B. 1 and 3
  C. 2 and 4
  D. only 4
  E. 1, 2, 3, and 4

7. The p-value is
A. the probability of the null hypothesis being true
B. the probability of the null hypothesis being false
C. the probability of the test statistic or any more extreme result, assuming the null hypothesis is true
D. the probability of the test statistic or any more extreme result, assuming the null hypothesis is false
E. none of the above

8. A clinical trial is conducted to compare the efficacy of two treatments, A and B. The difference between the mean effects of the two treatments is not statistically significant. This failure to reject the null hypothesis could be because
(1) the sample size is small
(2) the power of the statistical test is small
(3) the difference between the therapies is small
(4) the common variance is small
  A. 1, 2, and 3
  B. 1 and 3

C. 2 and 4
D. only 4
E. 1, 2, 3, and 4

9. A type II error is
A. the probability that the null hypothesis is true
B. the probability that the null hypothesis is false
C. made if the null hypothesis is accepted when it is false
D. made if the null hypothesis is rejected when it is true
E. none of the above

10. The power of a statistical test
(1) should be investigated whenever a significant result is obtained
(2) is a measure of significance
(3) increases with the variance of the population
(4) depends upon the sample size
A. 1, 2, and 3
B. 1 and 3
C. 2 and 4
D. only 4
E. 1, 2, 3, and 4

## CHAPTER 8

# The Many Uses of Chi-Square

**Key Concepts:** chi-square distribution; observed and expected values, goodness-of-fit tests; contingency table, dependent (response) and independent (factor) variables, association; McNemar's test for matched pairs; combining p-values; likelihood ratio criterion.

## THE CHI-SQUARE DISTRIBUTION

In this chapter we introduce a distribution known as chi-square, denoted $\chi^2$ ($\chi$ is the Greek letter chi, pronounced as "kite" without the "t" sound). This distribution has many important uses, including testing hypotheses about proportions and calculating confidence intervals for variances. Often you will read in the medical literature that "the chi-square test" was performed, or "chi-square analysis" was used, as though referring to a unique procedure. We shall see in this chapter that chi-square is not just a single method of analysis, but rather a distribution that is used in many different statistical procedures.

Suppose a random variable Y is known to have a normal distribution with mean $\mu$ and variance $\sigma^2$. We have stated that under these circumstances,

$$Z = \frac{Y - \mu}{\sigma}$$

has a standardized normal distribution (i.e., a normal distribution with mean zero and standard deviation one). Now the new random variable obtained by squaring Z, that is,

$$Z^2 = \frac{(Y - \mu)^2}{\sigma^2},$$

has a chi-square distribution with one degree of freedom. Only one random variable, Y, is involved in $Z^2$, and there are no constraints; hence we say $Z^2$ has one degree of freedom. From now on we shall abbreviate "degrees of freedom" by d.f.: the distribution of $Z^2$ is chi-square with 1 d.f.

Whereas a normally distributed random variable can take on any value, positive or negative, a chi-square random variable can only be positive or zero (a squared number cannot be negative). Recall that about 68% of the distribution of Z lies between $-1$ and $+1$; correspondingly, 68% of the distribution of $Z^2$, the chi-square distribution with 1 d.f., lies between 0 and $+1$. The remaining 32% lies between $+1$ and $\infty$. Therefore, the graph of $\chi^2$ with 1 d.f. is positively skewed, as shown in Figure 8.1.

Suppose that $Y_1$ and $Y_2$ are independent random variables, each normally distributed with mean $\mu$ and variance $\sigma^2$. Then

$$Z_1^2 = \frac{(Y_1 - \mu)^2}{\sigma^2}$$

and

$$Z_2^2 = \frac{(Y_2 - \mu)^2}{\sigma^2}$$

are each distributed as $\chi^2$ with one d.f. Moreover, $Z_1^2 + Z_2^2$ (i.e., the sum of these two independent, squared standardized normal random variables) is distributed as chi-square with 2 d.f. More generally, if we have a set of k independent random variables $Y_1, Y_2, \ldots, Y_k$, each normally distributed with mean $\mu$ and variance $\sigma^2$, then

$$\frac{(Y_1 - \mu)^2}{\sigma^2} + \frac{(Y_2 - \mu)^2}{\sigma^2} + \ldots + \frac{(Y_k - \mu)^2}{\sigma^2}$$

is distributed as $\chi^2$ with k degrees of freedom.

Now consider replacing $\mu$ by its minimum variance unbiased estimator $\overline{Y}$, the sample mean. Once the sample mean $\overline{Y}$ is determined, there are $k - 1$ choices possible for the values of the Y's. Thus

$$\frac{(Y_1 - \overline{Y})^2}{\sigma^2} + \frac{(Y_2 - \overline{Y})^2}{\sigma^2} + \ldots + \frac{(Y_k - \overline{Y})^2}{\sigma^2}$$

is distributed as $\chi^2$ with $k - 1$ degrees of freedom.

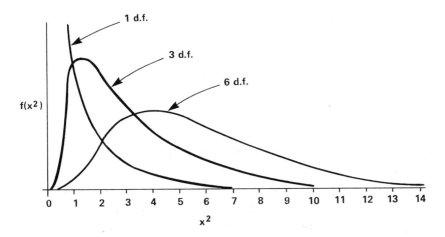

FIGURE 8.1 The chi-square distributions with 1, 3, and 6 d.f.

Figure 8.1 gives examples of what the chi-square distribution looks like. It is useful to remember that the mean of a chi-square distribution is equal to its number of d.f., and that its variance is equal to twice its number of d.f. Note also that as the number of d.f. increases, the distribution becomes less and less skewed; in fact, as a number of d.f. tends to infinity, the chi-square distribution tends to a normal distribution.

We now discuss some random variables that have approximate chi-square distributions. Let Y have an arbitrary distribution with mean $\mu$ and variance $\sigma^2$. Let $\overline{Y}$ be the mean of a random sample from the distribution. We know that for large samples

$$Z = \frac{\overline{Y} - \mu}{\sigma_{\overline{Y}}}$$

is approximately distributed as the standard normal, regardless of the shape of the distribution of Y. It follows that for large samples

$$Z^2 = \frac{(\overline{Y} - \mu)^2}{\sigma_{\overline{Y}}^2}$$

is approximately distributed as chi-square with 1 d.f., regardless of the shape of the distribution of Y. Similarly, if we sum k such quantities—each being the square of a standardized sample mean— then, provided they are independent, the sum is approximately distributed as chi-square, with k degrees of freedom.

Recall, for example, that if y is the outcome of a binomial random variable with parameters n and $\pi$, then in large samples the standardized variable

$$z = \frac{y/n - \pi}{\sqrt{\pi(1-\pi)/n}} = \frac{y - n\pi}{\sqrt{n\pi(1-\pi)}}$$

can be considered as the outcome of a standard normal random variable. Thus the square of this, $z^2$, can be considered to come from a chi-square distribution with 1 d.f. Now instead of writing y and n − y for the numbers we observe in the two categories (e.g., the number affected and the number unaffected), let us write $y_1$ and $y_2$, so that $y_1 + y_2 = n$. Analogously, let us write $\pi_1$ and $\pi_2$ so that $\pi_1 + \pi_2 = 1$. Then

$$z^2 = \frac{(y_1 - n\pi_1)^2}{n\pi_1(1 - \pi_1)} = \frac{(y_1 - n\pi_1)^2}{n\pi_1} + \frac{(y_2 - n\pi_2)^2}{n\pi_2}$$

(If you wish to follow the steps that show this, see Appendix 2). Now notice that each of the two terms on the right corresponds to one of the two possible outcomes for a binomially distributed random variable: $y_1$ and $n\pi_1$ are the observed and expected numbers in the first category (affected), and $y_2$ and $n\pi_2$ are the observed and expected numbers in the second category (unaffected). Each term is of the form

$$\frac{(\text{observed} - \text{expected})^2}{\text{expected}}$$

Adding them together, we obtain a statistic that comes from (approximately, in large samples) a chi-square distribution with 1 d.f. We shall see how this result can be generalized to generate many different "chi-square statistics," often called "Pearson chi-squares" after the British statistician Karl Pearson (1857–1936).

## GOODNESS-OF-FIT TESTS

To illustrate the use of this statistic, consider the offspring of matings in which one parent is hypercholesterolemic and the other is normocholesterolemic. We wish to test the hypothesis that children from such matings are observed in the 1:1 ratio of hypercholesterolemic to normocholesterolemic, as expected from matings of this type if hypercholesterolemia is a rare autosomal dominant trait. Thus we observe a set of children from such matings and wish to test the "goodness of fit" of the data to the hypothesis of autosomal dominant inheritance. A 1:1 ratio implies probabilities $\pi_1 = \pi_2 = 0.5$ for each category, and this will be our null hypothesis, $H_0$. Suppose the observed numbers are 87 and 79, so that n = 87 + 79 = 166. The expected numbers under $H_0$ are $n\pi_1 = 166(0.5) = 83$, and $n\pi_2 = 166(0.5) = 83$. The chi-square statistic is thus

$$\chi^2 = \frac{(87 - 83)^2}{83} + \frac{(79 - 83)^2}{83} = 0.39.$$

If we had observed the number of children expected under $H_0$ in each category (i.e., 83), the chi-square statistic, $\chi^2$, would be zero. Departure from this in either direction (either too many hypercholesterolemic or too many normocholesterolemic offspring) increases $\chi^2$. Thus we reject $H_0$ if $\chi^2$ is large (i.e., above the 95th percentile of the chi-square distribution with 1 d.f. for a test at the 5% significance level). The p-value is the area of the chi-square distribution above the observed $\chi^2$, and this automatically allows for departure from $H_0$ in either direction. From Table 3 in Appendix 2, we see that the 95th percentile of the chi-square distribution with 1 d.f. is 3.84. Since 0.39 is less than 3.84, the departure from $H_0$ is not significant at the 5% level. In fact 0.39 corresponds to p = 0.54, so that the fit to autosomal dominant inheritance is very good.

This test can easily be generalized to any number of categories. Suppose we have a sample of n observations, and each observation must fall in one, and only one, of k possible categories. Denote the numbers that are observed in each category $o_1, o_2, \ldots o_k$, and the corresponding numbers that are expected (under a particular $H_0$) $e_1, e_2, \ldots, e_k$. Then the chi-square statistic is simply

$$\chi^2 = \frac{(o_1 - e_1)^2}{e_1} + \frac{(o_2 - e_2)^2}{e_2} + \ldots + \frac{(o_k - e_k)^2}{e_k},$$

and under $H_0$, this can be considered as coming from a chi-square distribution with k − 1 d.f. (Once the total number of observations, n, is fixed, arbitrary numbers can be placed in only k − 1 categories). Of course, the sample size must be large enough. The same rule of thumb that we have introduced before can be used to check this: if each expected value is at least 5, the chi-square approximation is good. The approximation may still be good if a few of the expected values are less than 5, but in that situation it is common practice to pool two or more of the categories with small expected values.

As an example with three categories, consider the offspring of parents whose red cells agglutinate when mixed with either anti-M or anti-N sera. If these reactions are detecting two alleles at a single locus, then the parents are heterozygous (MN). Furthermore, the children should be MM (i.e., their red cells agglutinate only with anti-M) with probability 0.25, MN (like their parents) with probability 0.5, or NN (i.e., their cells agglutinate only with anti-N) with probability 0.25. Suppose we test the bloods of 200 children and observe in the three categories: $o_1 = 42$, $o_2 = 106$, and $o_3 = 52$, respectively. To test how well these data fit the hypothesis of two alleles segregating at a single locus, we calculate the appropriate chi-square statistic, with $e_1 = 200(0.25) = 50$, $e_2 = 200(0.5) = 100$, and $e_3 = 200(0.25) = 50$. The computations can be conveniently arranged as in Table 8.1. In this case we compare 1.72 to the chi-square distribution with 2 ($k - 1 = 3 - 1 = 2$) d.f. Table 3 in Appendix 2 shows that the 95th percentile of the distribution is 5.99, and so the departure from what is expected is not significant at the 5% level. In fact $p = 0.43$, and once again the fit is good.

The same kind of test can be used to determine the goodness-of-fit of a set of sample data to any distribution. We mentioned in the last chapter that there are tests to determine if a set of data could reasonably come from a normal distribution, and this is one such test. Sup-

TABLE 8.1. Computation of the Chi-Square Statistic to Test Whether the MN Phenotypes of 200 Offspring of MN × MN Matings Are Consistent With a Two-Allele, One-Locus Hypothesis

| Hypothesized Genotype | Number Observed (o) | Number Expected (e) | (o−e) | Contribution to $\chi^2$ [$(o−e)^2/e$] |
|---|---|---|---|---|
| MM | 42 | 50 | −8 | 1.28 |
| MN | 106 | 100 | +6 | 0.36 |
| NN | 52 | 50 | +2 | 0.08 |
| Total | 200 | 200 | 0 | $\chi^2 = 1.72$ |

pose, for example, we wanted to test the goodness-of-fit of the serum cholesterol levels in Table 3.2 to a normal distribution. The table gives the observed numbers in 20 different categories. We obtain the expected numbers in each category on the basis of the best-fitting normal distribution, substituting the sample mean and variance for the population values. In this way we have 20 observed numbers and 20 expected numbers, and so can obtain $\chi^2$ as the sum of 20 components. Because we force the expected number to come from a distribution with exactly the same mean and variance as in the sample, however, there are two fewer degrees of freedom. Thus in this case we would compare $\chi^2$ to the chi-square distribution with $20 - 1 - 2 = 17$ d.f. Note, however, that in the extreme categories of Table 3.2 there are some small numbers. If any of these categories have expected numbers below five, it might be necessary to pool them with neighboring categories. The total number of categories, and hence also the number of d.f., would then be further reduced.

## CONTINGENCY TABLES

Categorical data are often arranged in a table of rows and columns in which the individual entries are counts. Such a table is called a contingency table. Two qualitative variables are involved, the rows representing the categories of the first variable and the columns representing the categories of the second variable. Each cell in the table represents a combination of categories of the two variables. Each entry in a cell is the number of study units observed in that combination of categories. Tables 8.2 and 8.5 are each examples of contingency tables with two rows and two columns (the row and column indicating totals are not counted).

TABLE 8.2. 215 Medical Students Classified by Class and Serum Cholesterol Level (Above or Below 210 mg/100 ml)

| | Cholesterol Level | | |
| Class | Normal | High | Total |
|---|---|---|---|
| First year | 75 | 35 | 110 |
| Fourth year | 95 | 10 | 105 |
| Total | 170 | 45 | 215 |

In any two-way table, the hypothesis of interest—and hence the choice of an appropriate test statistic—is related to the types of variables being studied. We distinguish between two types of variables: dependent and independent variables. A **dependent**, or **response**, variable is one for which the distribution of study units in the different categories is merely observed by the investigator. A dependent variable is also sometimes called a criterion variable. An **independent**, or **factor**, variable is one for which the investigator actively controls the distribution of study units in the different categories. An independent variable is also sometimes called a predictor variable. Notice that we are using the term "independent" with a different meaning from that used in our discussion of probability. Although it is less confusing to use the terms "response" and "factor" variables, the terminology "dependent" and "independent" to describe these two different types of variables is so widely used that we shall use it too.

We shall consider the following two kinds of contingency tables:

1. The two-sample, one-independent variable and one-dependent variable case; the rows will be categories of an independent variable, whereas the columns will be categories of a dependent variable.

2. The one-sample, two-dependent variables case; the rows will be categories of one dependent variable and the columns will be categories of a second dependent variable.

CASE 1

Consider the data in Table 8.2. Ignoring the row and column of totals, this table has four cells. It is known as a fourfold table, or a 2 × 2 contingency table. The investigator took one sample of 110 first-year students and a second sample of 105 fourth-year students; these are two categories of an independent variable, since the investigator had control over how many first and fourth year students were taken. A blood sample was drawn from each student and analyzed for cholesterol level; each student was then classified as having normal or high cholesterol level based on a prespecified cut-point. These are two categories of a dependent variable, since the investigator observed, rather than controlled, how many students fell in each category. Thus, in this example, student class is an independent variable and cholesterol level is a dependent variable.

**TABLE 8.3.** Probability Structure for the Data in Table 8.2

| Class | Cholesterol Level | | Total |
|---|---|---|---|
| | Normal | High | |
| First year | $\pi_{11}$ | $\pi_{12}$ | 1 |
| Fourth year | $\pi_{21}$ | $\pi_{22}$ | 1 |
| Total | $\pi_{11} + \pi_{21}$ | $\pi_{12} + \pi_{22}$ | 2 |

Bearing in mind the way the data were obtained, we can view them as having the underlying probability structure shown in Table 8.3. Note that the first subscript of each $\pi$ indicates the row it is in, and the second indicates the column it is in. Note also that $\pi_{11} + \pi_{12} = 1$ and $\pi_{21} + \pi_{22} = 1$.

Suppose we are interested in testing the null hypothesis $H_0$ that the proportion of the first year class with high cholesterol levels is the same as that of the fourth year class (i.e., $\pi_{12} = \pi_{22}$). (If this is true, it follows automatically that $\pi_{11} = \pi_{21}$ as well). Assuming $H_0$ is true, we can estimate the "expected" number in each cell from the overall proportion of students who have normal or high cholesterol levels (i.e., from the proportions $170/215$ and $45/215$, respectively; see the last line of Table 8.2). In other words, if $\pi_{11} = \pi_{21}$ (which we shall then call $\pi_1$), and $\pi_{12} = \pi_{22}$ (which we shall then call $\pi_2$), we can think of the 215 students as forming a single sample and the probability structure indicated in the "Total" row of Table 8.3 becomes

| | | |
|---|---|---|
| $\pi_1$ | $\pi_2$ | 1 |

instead of

| | | |
|---|---|---|
| $\pi_{11} + \pi_{21}$ | $\pi_{12} + \pi_{22}$ | 2 |

We estimate $\pi_1$ by $p_1 = 170/215$ and $\pi_2$ by $p_2 = 45/215$. Now denote the total number of first-year students $n_1$ and the total number of fourth year students $n_2$. Then the expected numbers in each cell of the table are (always using the same convention for the subscripts—first one indicates the row, second one indicates the column):

$$e_{11} = n_1 p_1 = 110(170/215) = 86.98$$
$$e_{12} = n_1 p_2 = 110(45/215) = 23.02$$
$$e_{21} = n_2 p_1 = 105(170/215) = 83.02$$
$$e_{22} = n_2 p_2 = 105(45/215) = 21.98$$

Note that the expected number in each cell is the product of the marginal totals corresponding to that cell, divided by the grand total. The chi-square statistic is then given by

$$\chi^2 = \frac{(o_{11} - e_{11})^2}{e_{11}} + \frac{(o_{12} - e_{12})^2}{e_{12}} + \frac{(o_{21} - e_{21})^2}{e_{21}} + \frac{(o_{22} - e_{22})^2}{e_{22}}$$

$$= \frac{(75 - 86.98)^2}{86.98} + \frac{(35 - 23.02)^2}{23.02} + \frac{(95 - 83.02)^2}{83.02} + \frac{(10 - 21.98)^2}{21.98}$$

$$= 16.13 .$$

The computations can be conveniently arranged in tabular form, as indicated in Table 8.4.

In this case the chi-square statistic equals (about) the square of a single standardized normal random variable, and so has 1 d.f. Intuitively, we can deduce the number of d.f. by noting that we used the marginal totals to estimate the expected numbers in each cell, so that we forced the marginal totals of the expected numbers to equal the marginal totals of the observed numbers. (Check that this is so.) Now if we fix all the marginal totals, how many of the cells of the 2 × 2 table can be filled in with arbitrary numbers? The answer is only one; once we fill a single cell of the 2 × 2 table with an arbitrary number, that number and the marginal totals completely determine the other three entries in the table. Thus there is only 1 d.f. Looking up the percentile values of the chi-square distribution with 1 d.f. (Table 2, Appendix 2), we find that the 99.9th percentile is 10.83. Since 16.13 is greater than this, the two proportions are significantly different at the 0.1 % level (i.e., $p < 0.001$). We conclude that the proportion with high cholesterol levels is significantly different

**TABLE 8.4.** Computation of the Chi-Square Statistic for the Data in Table 8.2

| Class | Cholesterol Level | Number Observed (o) | Number Expected (e) | o−e | Contribution to $\chi^2$ [(o−e)²/e] |
|---|---|---|---|---|---|
| First year | Normal | 75 | 86.98 | −11.98 | 1.65 |
| | High | 35 | 23.02 | 11.98 | 6.23 |
| Fourth year | Normal | 95 | 83.02 | 11.98 | 1.72 |
| | High | 10 | 21.98 | −11.98 | 6.58 |
| Total | | 215 | 215.00 | | $\chi^2 = 16.13$ |

for first-year and fourth-year students. Equivalently, we conclude that the distribution of cholesterol levels depends on (is associated with) the class to which a student belongs, or that the two variables student class and cholesterol level are not independent **in the probability sense.** (As indicated earlier, this statement has nothing to do with our calling student class an independent variable and cholesterol level a dependent variable).

CASE 2

Consider now the data given in Table 8.5. Here we have a single sample of 110 first-year medical students and have observed whether each student is high or normal, with respect to specified cut-points, for two dependent variables: cholesterol level and triglyceride level. These data can be viewed as having the underlying probability structure shown in Table 8.6, which should be contrasted with Table 8.3. Notice that dots are used in the marginal totals of Table 8.6 (e.g., $\pi_{1.} = \pi_{11} + \pi_{12}$); so that a dot replacing a subscript indicates that the $\pi$ is the sum of the $\pi$'s with different values of that subscript.

Suppose we are interested in testing the null hypothesis $H_0$ that triglyceride level is not associated with cholesterol level, (i.e., triglyceride level is independent of cholesterol level in a probability sense).

**TABLE 8.5.** 110 First-year Medical Students Classified by Serum Cholesterol Level (Above or Below 210 mg/100 ml) and Serum Triglyceride Level (Above or Below 150 mg/100 ml

|  | Triglyceride Level | | |
| --- | --- | --- | --- |
| Cholesterol Level | Normal | High | Total |
| Normal | 60 | 15 | 75 |
| High | 20 | 15 | 35 |
| Total | 80 | 30 | 110 |

**TABLE 8.6.** Probability Structure for the Data in Table 8.5

|  | Triglyceride Level | | |
| --- | --- | --- | --- |
| Cholesterol Level | Normal | High | Total |
| Normal | $\pi_{11}$ | $\pi_{12}$ | $\pi_{1.}$ |
| High | $\pi_{21}$ | $\pi_{22}$ | $\pi_{2.}$ |
| Total | $\pi_{.1}$ | $\pi_{.2}$ | 1 |

Recalling the definition of independence from Chapter 4, we can state $H_0$ as being equivalent to

$$\pi_{11} = \pi_{1.}\pi_{.1}$$
$$\pi_{12} = \pi_{1.}\pi_{.2}$$
$$\pi_{21} = \pi_{2.}\pi_{.1}$$
$$\pi_{22} = \pi_{2.}\pi_{.2}.$$

Assuming $H_0$ is true, we can once again estimate the expected number in each cell of the table. We first estimate the marginal proportions of the table. Using the letter p to denote an estimate of the probability, these are

$$p_{1.} = \frac{75}{110}$$
$$p_{2.} = \frac{35}{110}$$
$$p_{.1} = \frac{80}{110}$$
$$p_{.2} = \frac{30}{110}$$

Then each cell probability is estimated as a product of the two corresponding marginal probabilities (since if $H_0$ is true we have independence). Thus, letting n denote the total sample size, under $H_0$ the expected numbers are calculated to be

$$e_{11} = n\, p_{1.}p_{.1} = 110 \left(\frac{75}{110}\right)\left(\frac{80}{110}\right) = 54.55$$
$$e_{12} = n\, p_{1.}p_{.2} = 110 \left(\frac{75}{110}\right)\left(\frac{30}{110}\right) = 20.45$$
$$e_{21} = n\, p_{2.}p_{.1} = 110 \left(\frac{35}{110}\right)\left(\frac{80}{110}\right) = 25.45$$
$$e_{22} = n\, p_{2.}p_{.2} = 110 \left(\frac{35}{110}\right)\left(\frac{30}{110}\right) = 9.55$$

Note that after cancelling out 110, each expected number is once again the product of the two corresponding marginal totals divided

**TABLE 8.7.** Computation of the Chi-Square Statistic for the Data in Table 8.5

| Cholesterol Level | Triglyceride Level | Number Observed $(o)$ | Number Expected $(e)$ | $o-e$ | Contribution to $\chi^2$ $[(o-e)^2/e]$ |
|---|---|---|---|---|---|
| Normal | Normal | 60 | 54.55 | +5.45 | 0.54 |
|  | High | 15 | 20.45 | −5.45 | 1.45 |
| High | Normal | 20 | 25.45 | −5.45 | 1.17 |
|  | High | 15 | 9.55 | +5.45 | 3.11 |
| Total |  | 110 | 110.00 |  | $\chi^2 = 6.27$ |

by the grand total. Thus we can calculate the chi-square statistic in exactly the same manner as before, and once again the resulting chi-square has 1 d.f. Table 8.7 summarizes the calculations. The calculated value, 6.27, lies between the 95th and 99th percentiles (3.84 and 6.63, respectively: see Table 3, Appendix 2) of the chi-square distribution with 1 d.f. We therefore conclude that triglyceride levels and cholesterol levels are not independent but are associated $(0.01 < p < 0.05)$.

Suppose now that we ask a different question, again to be answered using the data in Table 8.5: Is the proportion of students with high cholesterol levels different from the proportion with high triglyceride levels? In other words, we ask whether the two dependent variables, dichotomized, follow the same binomial distribution. Our null hypothesis $H_0$ is that the two proportions are the same; i.e.,

$$\pi_{21} + \pi_{22} = \pi_{12} + \pi_{22},$$

which is equivalent to $\pi_{21} = \pi_{12}$.

A total of $20 + 15 = 35$ students are in the two corresponding cells, and under $H_0$ the expected number in each would be half this, that is,

$$e_{12} = e_{21} = \frac{1}{2}(o_{12} + o_{21}) = 35/2 = 17.5$$

The appropriate chi-square statistic to test this hypothesis is thus

$$\chi^2 = \frac{(o_{12} - e_{12})^2}{e_{12}} + \frac{(o_{21} - e_{21})^2}{e_{21}} = \frac{(20 - 17.5)2}{17.5}$$
$$+ \frac{(15 - 17.5)2}{17.5} = 0.71$$

The numbers in the other two cells of the table are not relevant to the question asked, and so the chi-square statistic for this situation is formally the same as the one we calculated earlier to test for Mendelian proportions among offspring of hypercholesterolemic by normocholesterolemic parents. Once again it has 1 d.f. and there is no significant difference at the 5% level (in fact, $p = 0.4$).

Regardless of whether or not it would make any sense, we cannot apply the probability structure in Table 8.6 to the data in Table 8.2 and analogously ask whether $\pi_{.1}$ and $\pi_{1.}$ are equal (i.e., is the proportion of fourth-year students equal to the proportion of students with high cholesterol?). The proportion of students who are fourth-year cannot be estimated from the data in Table 8.2, because we were told that the investigator took a sample of 110 first-year students and a sample of 105 fourth-year students. The investigator had complete control over how many students in each class came to be sampled, regardless of how many there happened to be. If, on the other hand, we had been told that a random sample of all first- and fourth-year students had been taken, and it was then observed that the sample contained 110 first-year and 105 fourth-year students, then student class would be a dependent variable and we could test the null hypothesis $\pi_{.1} = \pi_{1.}$.

You might think that any hypothesis of this nature is somewhat artificial; after all, whether or not the proportion of students with high cholesterol is equal to the proportion with high triglyceride is merely a reflection of the cut-points used for each variable. There is, however, a special situation where this kind of question, requiring the test we have just described (which is called McNemar's test), often occurs. Suppose we wish to know whether the proportion of men and women with high cholesterol levels is the same, for which we would naturally define "high" by the same cut-point in the two sexes. One way to do this would be to take two samples—one of men and one of women—and perform the first test we described for a $2 \times 2$ contingency table (Case 1). The situation would be analogous to that summarized in Table 8.2, the independent variable being gender rather than class. But cholesterol levels change with age. Unless we take the precaution of having the same age distribution in each sample, any difference that is found could be due to either the age difference or the sex difference between the two groups. For this reason it would be wise to take a sample of matched pairs, each pair consisting of a man and a woman of the same age. If we have n such pairs, we do not have 2n independent individuals, because of the matching. By considering each **pair** as a study unit, however, it is reasonable to suppose that we have n independent study units, with two different dependent variables measured on each—cholesterol level of the woman of the pair and cholesterol level of the man of the pair. We would then draw up a table similar to Table 8.5, but with each **pair** as a study unit. Thus corresponding to the 110 medical students, we would have n, the number of pairs; and the two dependent variables, instead of cholesterol and triglyceride level, would be cholesterol level in the woman of each pair and cholesterol level in the man of each pair. To test whether the proportion with high cholesterol is the same in men and women, we would now use McNemar's test, which uses the information in only those two cells of the 2 × 2 table that relate to untied pairs.

There are thus two different ways in which we could conduct a study to answer the same question: Is cholesterol level independent of sex? Because of the different ways in which the data are sampled, two different chi-square tests are necessary: the first is the usual contingency table chi-square test, the second is McNemar's test.

In general, a contingency table can have any number, r, of rows and any number, c, of columns, and the contingency table chi-square is used to test whether the row variable is independent of or associated with the column variable. The general procedure is to use the marginal totals to calculate an "expected" number for each cell, and then to sum the quantities (observed − expected)$^2$/expected for all r × c cells. Fixing the marginal totals, it is found that $(r − 1)(c − 1)$ cells can be filled in with arbitrary numbers, and so this chi-square has $(r − 1)(c − 1)$ degrees of freedom. When $r = c = 2$ (the 2 × 2 table), $(r − 1)(c − 1) = (2 − 1)(2 − 1) = 1$. For the resulting statistic to be distributed as chi-square under $H_0$, we must

1. have each study unit appearing in only one cell of the table
2. sum the contributions over all the cells, so that **all** the study units in the sample are included
3. have in each cell the count of a number of **independent** events
4. not have small expected values causing large contributions to the chi-square.

Note condition 3. Suppose our study units are children and these have been classified according to disease status. If disease status is in any way familial, then two children in the same family are not independent. Although condition 3 would be satisfied if the table contains only one child per family, it would not be satisfied if sets of brothers and sisters are included in the table. In such a situation the "chi-square" statistic would not be expected to follow a chi-square distribution.

Before leaving the subject of contingency tables, a cautionary note is in order regarding the interpretation of any significant dependence or association that is found. As stated in Chapter 4, there can be many different causes underlying the dependence between two events. Consider, for example, the following fourfold table, in which 2,000 persons are classified as possessing a particular antigen (A+) or not (A−), and as having a particular disease (D+) or not (D−):

|  | A+ | A− | Total |
|---|---|---|---|
| D+ | 51 | 59 | 110 |
| D− | 549 | 1,341 | 1,890 |
| Total | 600 | 1,400 | 2,000 |

We see that among those persons who have the antigen, $51/600 = 8.5\%$ have the disease, whereas among those who do not have the antigen, $59/1,400 = 4.2\%$ have the disease. There is a clear association between the two variables, which is highly significant (chi-square with 1 d.f. $= 14.84$, $p < 0.001$). Does this mean that possession of the antigen predisposes to having the disease? Or that having the disease predisposes to possession of the antigen? Neither of these interpretations may be correct, as we shall see.

Consider the following two analogous fourfold tables, one pertaining to 1,000 white persons and one pertaining to 1,000 black persons:

| | Whites | | | | | Blacks | | |
|---|---|---|---|---|---|---|---|---|
|  | A+ | A− | Total |  |  | A+ | A− | Total |
| D+ | 50 | 50 | 100 |  | D+ | 1 | 9 | 10 |
| D− | 450 | 450 | 900 |  | D− | 99 | 891 | 990 |
| Total | 500 | 500 | 1,000 |  | Total | 100 | 900 | 1,000 |

In neither table is there any association between possession of the antigen and having the disease. The disease occurs among 10 % of the white persons, whether or not they possess the antigen; and it occurs among 1 % of the black persons, whether or not they possess the antigen. The antigen is, however, more prevalent in the white sample than in the black sample. Because of this, when we add the two samples together—which results in the original table for all 2,000 persons—a significant association is found between possession of the antigen and having the disease. Thus an association can be caused merely by mixing samples from two or more populations, or by sampling from a single heterogeneous population. Populations may be heterogeneous with respect to race, age, sex, or any number of other factors that could be the cause of an association.

## INFERENCE ABOUT THE VARIANCE

Let us suppose we have a sample $Y_1, Y_2, \ldots, Y_n$ from a normal distribution with variance $\sigma^2$, and let the sample mean be $\overline{Y}$. We have seen that

$$\frac{(Y_1 - \overline{Y})^2}{\sigma^2} + \frac{(Y_2 - \overline{Y})^2}{\sigma^2} + \ldots + \frac{(Y_n - \overline{Y})^2}{\sigma^2}$$

then follows a chi-square distribution with $n-1$ degrees of freedom. But this expression can also be written in the form

$$\frac{(n - 1)S^2}{\sigma^2},$$

where $S^2$ is the sample variance. Thus, denoting the 2.5th and 97.5th percentiles of the chi-square distribution with $n - 1$ d.f. as $\chi^2_{2.5}$ and $\chi^2_{97.5}$, respectively, we know that

$$P(\chi^2_{2.5} \leq \frac{(n-1)S^2}{\sigma^2} \leq \chi^2_{97.5}) = 0.95 \,.$$

This statement can be written in the equivalent form

$$P\left(\frac{(n-1)S^2}{\chi^2_{97.5}} \leq \sigma^2 \leq \frac{(n-1)S^2}{\chi^2_{2.5}}\right) = 0.95 \,,$$

which gives us a way of obtaining a 95% confidence interval for a variance; all we need do is substitute the specific numerical value $s^2$ from our sample in place of the random variable $S^2$. In other words, we have 95% confidence that the true variance $\sigma^2$ lies between the two numbers

$$\frac{(n-1)s^2}{\chi^2_{97.5}} \text{ and } \frac{(n-1)s^2}{\chi^2_{2.5}} .$$

For example, suppose we found $s^2 = 4$ with 10 d.f. Table 3 in Appendix 2 does not give the 2.5th and 97.5th percentiles, so we shall use the 5th and 95th percentiles of the chi-square distribution with 10 d.f. (3.94 and 18.31) to obtain a 90% confidence interval. The limits of the interval are

$$\frac{(10)4}{18.31} = 2.18 \text{ and } \frac{(10)4}{3.94} = 10.15 .$$

Notice that even though this is a 90% confidence interval (which is necessarily narrower than a 95% confidence interval), it is nevertheless quite wide. Typically, we require a much larger sample to estimate a variance or standard deviation than we require to estimate, with the same degree of precision, a mean.

We can also test hypotheses about a variance. If we wanted to test the hypothesis $\sigma^2 = 6$ in the above example, we would compute

$$\chi^2 = \frac{(10)(4)}{6} = 6.67$$

which, if the hypothesis is true, comes from a chi-square distribution with 10 degrees of freedom. Since 6.67 is between the 70th and 80th percentiles of that distribution, there is no evidence to reject the hypothesis.

We have already discussed the circumstances under which the F-test can be used to test the hypothesis that two population variances are equal. Although the details are beyond the scope of this book, you should be aware of the fact that it is possible to test for the equality of a set of more than two variances, and that at least one of the tests to do this is based on the chi-square distribution. Remember, however, that all chi-square procedures for making inferences about variances depend rather strongly on the assumption of normality; they are quite sensitive to non-normality of the underlying variable.

# COMBINING P-VALUES

Suppose five investigators have conducted different experiments to test the same null hypothesis $H_0$ (for example, that two treatments have the same effect). Suppose further that the tests of significance of $H_0$ (that the mean response to treatment is the same) resulted in the p-values $p = 0.15$, $p = 0.07$, $p = 0.50$, $p = 0.22$, and $p = 0.09$. At first glance you might conclude that there is no significant difference between the two treatments. There is a way of pooling p-values from separate investigations, however, to obtain an overall p-value.

For any arbitrary p-value, if the null hypothesis that gives rise to it is true, $-2 \log_e p$ can be considered as coming from the $\chi^2$ distribution with 2 d.f. ($\log_e$ stands for "logarithm to base e," or natural logarithm; it is denoted ln on many calculators). If there are k independent investigations, the corresponding p-values will be independent. Thus the sum of k such values,

$$ -2 \log_e p_1 - 2 \log_e p_2 - \ldots - 2 \log_e p_k \, , $$

can be considered as coming from the $\chi^2$ distribution with 2k d.f. Hence, in the above example, we would calculate

$$
\begin{aligned}
-2 \log_e (0.15) &= 3.79 \\
-2 \log_e (0.07) &= 5.32 \\
-2 \log_e (0.50) &= 1.39 \\
-2 \log_e (0.22) &= 3.03 \\
-2 \log_e (0.09) &= \underline{4.82} \\
\text{Total} \quad &= 18.35
\end{aligned}
$$

If $H_0$ is true, 18.35 comes from a $\chi^2$ distribution with $2k = 2(5) = 10$ d.f. From Table 3, Appendix 2, we see that for the distribution with 10 d.f., 18.31 corresponds to $p = 0.05$. Thus, by pooling the results of all five investigations, we see that the treatment difference is just significant at the 5% level. It is, of course, necessary to check that each investigator is in fact testing the same null hypothesis.

## LIKELIHOOD RATIO TESTS

Earlier we introduced maximum likelihood estimators as a type of estimator with many good properties. Since different statistical tests compete with each other in much the same way that different estimators do, it would be helpful to have a general approach to derive statistical tests that have desirable properties. One such approach is to use a test based on the **likelihood ratio criterion.**

Recall that we ended Chapter 4 by calling the ratio of two conditional probabilities a likelihood ratio. In that instance, we were discussing the likelihood of a child receiving a B allele conditional on the child's father being a man accused of being the father, relative to the likelihood of the same event conditional on the child's father being a random man from a particular population. We can think of these two possibilities as two different hypotheses. In general, the likelihood ratio criterion is simply the likelihood of a set of data under one hypothesis divided by the likelihood of the same set of data under a second hypothesis. In either case, however, the likelihood may depend upon unknown parameters. When this occurs we simply substitute the maximum likelihood estimates for the parameters. In other words, each likelihood is maximized (mathematically or numerically) over any unknown parameters.

Suppose, for example, we have two random samples. Under one hypothesis (the null hypothesis) they both come from identical normal distributions — same means and variances; under the other, alternative, hypothesis they come from two normal distributions with the same variances but different means. We wish to test the null hypothesis that the means of the two samples are not different. (Of course, we already know that the two-sample t-test is appropriate for this situation, but we shall nevertheless use it to illustrate how the likelihood ratio test works). Under each hypothesis the likelihood depends on the mean(s) and the variance, which are unknown parameters. So, to obtain the likelihood ratio, we maximize the likelihood twice: once over all possible values for a common mean and a common variance (i.e., under the null hypothesis), and once over all possible values for separate sample means and a common variance (the alternative hypothesis). The likelihood ratio criterion is then the maximum likelihood under the null hypothesis divided by the maximum likelihood under the alternative hypothesis.

The beauty of the likelihood ratio criterion lies in the following general theorem that holds under certain well-defined conditions: As the sample size increases, $- 2 \log_e$ (likelihood ratio), i.e., minus twice the natural logarithm of the likelihood ratio, tends to be distributed as chi-square if the null hypothesis is true. Thus, given large samples, if the two population means in our example are identical, $- 2 \log_e$ (likelihood ratio) is approximately distributed as chi-square. The number of d.f. is equal to the number of restrictions implied by the null hypothesis. In our example, the null hypothesis is that the two means are equal, which is a single restriction, so there is 1 d.f. For this particular example the test becomes identical to the two-sample t-test as the sample size tends to infinity. In fact, most of the tests discussed in this book are "asymptotically" identical to a test based on the likelihood ratio criterion. In those cases in which it has not been mathematically possible to derive an exact test, this general test based on a chi-square distribution is often used. Since it is now feasible, with modern computers, to calculate likelihoods corresponding to very elaborate probability models, this general approach is becoming more common in the medical literature. We shall discuss an example in Chapter 11.

## SUMMARY

1. Chi-square is a family of distributions used in many statistical procedures. Theoretically, the chi-square distribution with k d.f. is the distribution of the sum of k independent random variables, each of which is the square of a standardized normal random variable.

2. In practice we often sum more than k quantities that are not independent, but the sum is in fact equivalent to the sum of k independent quantities. The integer k is the number of degrees of freedom (d.f.) associated with the chi-square distribution. In most situations there is an intuitive way of determining the number of d.f. When the data are counts, we often sum quantities of the form (observed−expected)$^2$/expected; the number of d.f. is then the number of counts that could have been arbitrarily chosen—with the stipulation that there is no change in the total number of counts or other specified parameters. Large values of the chi-square statistic indicate departure from the null hypothesis.

3. A chi-square goodness-of-fit test can be used to test whether a sample of data is consistent with any specified probability distribution. In the case of continuous traits, the data are first categorized in

the manner used to construct a histogram. Categories with small expected numbers (less than 5) are usually pooled into larger categories.

4. In a two-way contingency table, either or both of the row and column variables may be dependent (response) variables. One variable may be controlled by the investigator, and is then called an independent (factor) variable.

5. The hypothesis of interest determines which chi-square test is performed. Association, or lack of independence between two variables, is tested by the usual contingency table chi-square. The expected number in each cell is obtained as the product of the corresponding row and column totals divided by the grand total. The number of degrees of freedom is equal to the product: (number of rows − 1)(number of columns −1). Each study unit must appear only once in the table, and each count within a cell must be the count of a number of independent events.

6. For a 2 × 2 table in which both rows and columns are dependent variables, McNemar's test is used to test whether the two variables follow the same binomial distribution. If the study units are matched pairs (e.g., men and women matched by age), and each variable is a particular measurement on a specific member of the pair (e.g., cholesterol level on the man of the pair and the cholesterol level on the woman of the pair), then McNemar's test is used to test whether the binomial distribution (low or high cholesterol level) is the same for the two members of the pair (men and women). This tests whether the specific measurement (cholesterol level) is independent of the member of the pair (gender).

7. The chi-square distribution can be used to construct a confidence interval for the variance of a normal random variable, or to test that a variance is equal to a specified quantity. This interval and this test are not robust against non-normality.

8. A set of p-values that result from independent investigations, all testing the same null hypothesis, can be combined to give an overall test of the null hypothesis. The sum of k independent quantities, $-2 \log_e p$, is compared to the chi-square distribution with 2k d.f.; a significantly large chi-square suggests that overall, the null hypothesis is not true.

9. The likelihood ratio criterion provides a general method of testing a hypothesis in large samples. Many of the usual statistical tests

become identical to a test based on the likelihood ratio criterion as the sample size becomes infinite. The likelihood ratio criterion is the maximum likelihood of a particular sample of data under one (the null) hypothesis divided by the maximum likelihood of the same sample of data under another (alternative) hypothesis. Under certain well defined conditions, $- 2 \log_e$ (likelihood ratio) can be used to test the null hypothesis, for it is approximately distributed as chi-square. The number of d.f. is the same as the number of restrictions implied by the null hypothesis.

## PROBLEMS

1. The chi-square distribution is useful in all the following except
   A. testing the equality of two proportions
   B. combining a set of three p-values
   C. testing for association in a contingency table
   D. testing the hypothesis that the variance is equal to a specific value
   E. testing the hypothesis that two variances are equal

2. An investigator is studying the response to treatment for angina. Patients were randomly assigned to one of two treatments, and each patient's response was recorded in one of four categories. An appropriate test for the hypothesis of equal response patterns for the two treatments is the
   A. t-test
   B. F-test
   C. z-test
   D. $\chi^2$-test
   E. rank sum test

3. For Problem 2, the appropriate number of degrees of freedom is
   A. 1
   B. 2
   C. 3
   D. 4
   E. 5

4. Blood samples were taken from a sample of 100 medical students and serum cholesterol levels determined. A histogram suggested

the serum cholesterol levels are approximately normally distributed. A chi-square goodness-of-fit test for normality yielded $\chi^2 = 9.05$ with 12 d.f. ($p = 0.70$). An appropriate conclusion is:
   A. the data are consistent with the hypothesis that their distribution is normal
   B. the histogram is misleading in situations like this; a Poisson distribution would be more appropriate
   C. the goodness-of-fit test cannot be used for testing normality
   D. a scatter diagram should have been used to formulate the hypothesis
   E. none of the above

5. An investigator is studying the association between dietary and exercise habits in a group of 300 students. He summarizes the findings as follows:

| Dietary Habits | Exercise Habits | Number Observed (O) | Number Expected (E) | O−E | Contribution to $\chi^2$ |
|---|---|---|---|---|---|
| Poor | Poor | 23 | 27.45 | −4.45 | 0.72 |
| | Moderate | 81 | 68.85 | 12.15 | 2.14 |
| | Good | 31 | 38.70 | −7.70 | 1.53 |
| Moderate | Poor | 15 | 17.08 | −2.08 | 0.25 |
| | Moderate | 47 | 42.84 | 4.16 | 0.40 |
| | Good | 22 | 24.08 | −2.08 | 0.18 |
| Good | Poor | 23 | 16.47 | 6.53 | 2.59 |
| | Moderate | 35 | 41.31 | −16.31 | 6.44 |
| | Good | 33 | 23.22 | 9.78 | 4.12 |
| Total | | 300 | 300.00 | | $\chi^2 = 18.37$ |

The chi-square test with 4 d.f. was reported to be statistically significant ($p = 0.002$). Which of the following is true?
(1) The correct number of degrees of freedom is not 4.
(2) The correct number of degrees of freedom is 8.
(3) The chi-square is smaller than expected if there is no association.
(4) The data are inconsistent with the hypothesis of no association.
   A. 1, 2, and 3
   B. 1 and 3
   C. 2 and 4
   D. Only 4
   E. 1, 2, 3, and 4

6. Two drugs—an active compound and a placebo—were compared for their ability to relieve anxiety. Sixty patients were randomly assigned to one or the other of the two treatments. After 30 days on treatment the patients were evaluated in terms of improvement or no improvement. The study was double blind. A chi-square test was performed to compare the proportions of improved patients, resulting in $\chi^2 = 7.91$ with 1 d.f. (p $= 0.005$). A larger proportion improved in the active compound group. An appropriate conclusion is
   A. the placebo group was handicapped by the random assignment to groups
   B. an F-test is needed to evaluate the data
   C. the data suggest the two treatments are approximately equally effective in relieving anxiety
   D. the data suggest the active compound is more effective than placebo in relieving anxiety
   E. none of the above

7. A lipid laboratory claimed it could determine serum cholesterol levels with a standard deviation less than 5 mg/100 ml. Samples of blood were taken from a series of patients. The blood was pooled, thoroughly mixed, and divided into aliquots. Ten of these aliquots were labeled with fictitious names and sent to the lipid laboratory for routine lipid analysis, interspersed with blood samples from other patients. Thus, the cholesterol determinations for these aliquots should have been identical, except for laboratory error. On examination of the data, the standard deviation of the 10 aliquots was found to be 7 mg/100 ml. Assuming cholesterol levels are approximately normally distributed, a chi-square test was performed of the null hypothesis that the standard deviation is 5; it was found that chi-square $= 17.64$ with 9 d.f. (p $= 0.04$). An appropriate conclusion is
   A. the data are consistent with the laboratory's claim
   B. the data suggest the laboratory's claim is not valid
   C. rather than a $\chi^2$-test, a t-test is needed to evaluate the claim
   D. the data fail to shed light on the validity of the claim
   E. a clinical trial would be more appropriate for evaluating the claim

8. A group of 180 students were interviewed to see how many follow a prudent diet. They were then given a 90-day series of in-

depth lectures, including clinical evaluations on nutrition and its association with heart disease and cancer. One year later the students were re-interviewed and assessed for the type of diet they followed, yielding the following data:

| Prudent Diet Initially | Prudent Diet at Follow-up | | Total |
|---|---|---|---|
| | Yes | No | |
| Yes | 21 | 17 | 38 |
| No | 37 | 105 | 142 |
| Total | 58 | 122 | 180 |

McNemar's test results in $\chi^2 = 8.96$ with 1 d.f. (p = 0.004). An appropriate conclusion is
  A. the study is invalid since randomization was not used
  B. the effect of the lectures is confounded with that of the initial weight of the students
  C. the data suggest the lectures were ineffective
  D. the lectures appear to have had an effect
  E. none of the above

9. An investigator performed an experiment to compare two treatments for a particular disease. He analyzed the results using a t-test and found p = 0.08. Since he had decided to declare the difference statistically significant only if p < 0.05, he decided his data were consistent with the null hypothesis. Several days later he discovered a paper on a similar previous study which reported p = 0.11. Further review of the literature produced two additional studies with p-values 0.19 and 0.07. Since the treatment differences were in the same direction in all four studies, the investigator computed

$$\chi^2 = -2(\log_e 0.08 + \log_e 0.11 + \log_e 0.19 + \log_e 0.07)$$
$$= 18.10 \text{ with 8 d.f. (p = 0.013).}$$

An appropriate conclusion is
  A. the investigator should not combine p-values from different studies
  B. although none of the separate p-values is significant at the 0.05 level, the combined value is

C. the t-test should be used to combine p-values
D. the combined p-value is not statistically significant
E. the number of p-values combined is insufficient to warrant making a decision

10. In a case-control study, the proportion of cases exposed to a suspected carcinogen is reported to be **not** significantly different from the proportion of controls exposed (chi-square with 1 d.f. = 1.33, p = 0.25). A 95% confidence interval for the odds ratio for these data is reported to be 2.8 ± 1.2. An appropriate conclusion is

A. there is no evidence that the suspected carcinogen is related to the risk of being a case
B. the reported results are inconsistent, and therefore no conclusion can be made
C. the p-value is such that the results should be declared statistically significant
D. we cannot study the effect of the suspected carcinogen in a case-control study
E. none of the above

## CHAPTER 9

# Correlation and Regression

**Key Concepts:** linear regression: slope, intercept, residual, error or residual sum of squares, sum of squares due to regression, mean squares, error mean square; homoscedasticity, heteroscedasticity; linear relationship, covariance, product-moment and rank correlation; multiple regression, stepwise regression, multiple correlation coefficient, partial correlation coefficient; regression toward the mean

## SIMPLE LINEAR REGRESSION

In Chapter 8 we discussed categorical data and introduced the notion of dependent and independent variables. We turn now to a discussion of relationships between dependent and independent variables of the continuous type. To explain such a relationship we often search for mathematical models such as equations for straight lines, parabolas, or other mathematical functions. The anthropologist Sir Francis Galton (1822–1911) used the term **regression** in explaining the relationship between the heights of fathers and their sons. From a group of 1,078 father-son pairs he developed the following model, in which the heights are in inches:

son's height $= 33.73 + 0.516$ (father's height)

If we substitute 74 inches into this equation for the father's height, we arrive at about 72 inches for the son's height (i.e., the son is not as tall as his father). On the other hand, if we substitute 65 inches for the father's height, we find the son's height to be 67 inches (i.e., the son is taller than his father). Galton concluded that although tall fathers tend to have tall sons and short fathers tend to have short sons, the son's height tends to be closer to the average than his father's height.

Galton called this "regression toward the mean." Although the techniques for modeling relationships among variables have taken on a much wider meaning, the term "regression" has become entrenched in the statistical literature to describe this modeling. Thus, nowadays we speak of regression models, regression equations, and regression analysis without wishing to imply that there is anything "regressive." As we shall see later, however, the phrase "regression toward the mean" is still used in a sense analogous to what Galton meant by it.

We begin by discussing a relatively simple relationship between two variables, namely a straight-line relationship. The equation $y = 3 + 2x$ is the equation of the straight line shown in Figure 9.1. The equation $y = 2 - 3x$ is shown in Figure 9.2, and the equation $y = 0.5x$ in Figure 9.3.

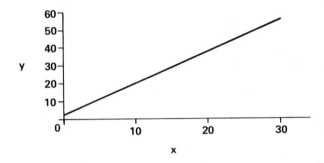

**FIGURE 9.1** Graph of the equation $y = 3 + 2x$.

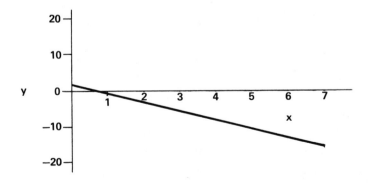

**FIGURE 9.2** Graph of the equation $y = 2 - 3x$.

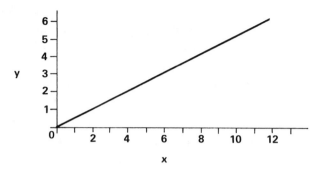

FIGURE 9.3 Graph of the equation y = 0.5x.

In general, the equation of a straight line can be expressed in the form

$$y = \beta_0 + \beta_1 x$$

where $\beta_0$ and $\beta_1$ are specified constants. Any point on this line has an x-coordinate and a y-coordinate. When $x = 0$, $y = \beta_0$; so the parameter $\beta_0$ is the value of the y-coordinate where the line crosses the y-axis, and is called the **y-intercept**. In Figure 9.1 the intercept is 3, in Figure 9.2 it is 2. When $\beta_0 = 0$ the line goes through the origin (Figure 9.3) and the intercept is 0. The parameter $\beta_1$ is the **slope** of the line and measures the amount of change in y per unit increase in x. When $\beta_1$ is positive (Figures 9.1 and 9.3), the slope is upward and x and y increase together; when $\beta_1$ is negative (Figure 9.2), the slope is downward and y decreases as x increases. When $\beta_1$ is zero, y is the same (it has value $\beta_0$) for all values of x and the line is horizontal (i.e., there is no slope). The parameter $\beta_1$ is undefined for vertical lines but approaches infinity as the line approaches a vertical position.

So far in our discussion we have assumed that the relationship between x and y is explained exactly by a straight line; if we are given x we can determine y—and vice versa—for all values of x and y. Now let us assume that the relationship between the two variables is not exact, because one of the variables is subject to random measurement errors. Let us call this random variable the dependent variable and denote it Y. The other variable x is assumed to be measured without error; it is under the control of the investigator and we call it the independent variable. This terminology is consistent with that of

Chapter 8. In practice the independent variable will also often be subject to random variability caused by errors of measurement, but we assume that this variability is negligible relative to that of the dependent variable. For example, suppose an investigator is interested in the rate at which a metabolite is consumed or produced by an enzyme reaction. A reaction mixture is set up from which aliquots are withdrawn at various intervals of time and the concentration of the metabolite in each aliquot determined. Whereas the time at which each aliquot is withdrawn can be accurately measured, the metabolite concentration, being determined by a rather complex assay procedure, is subject to appreciable measurement error. In this situation time would be the independent variable under the investigator's control, and metabolite concentration would be the random dependent variable.

Since the relationship is not exact, we write

$$Y = \beta_0 + \beta_1 x + \epsilon,$$

where $\epsilon$, called the **error**, is the amount by which the random variable Y, for a given x, lies above the line (or below the line, if it is negative). This is illustrated in Figure 9.4. In a practical situation we would have a sample of pairs of numbers, $x_1$ and $y_1$, $x_2$ and $y_2$, and so on. Then, assuming a straight line is an appropriate model, we would try to find the line that best fits the data. In other words, we would try to find estimates $b_0$ and $b_1$ of the parameters $\beta_0$ and $\beta_1$, respectively. One approach that yields estimates with good properties is to take the line that minimizes the sum of the squared errors (i.e., that

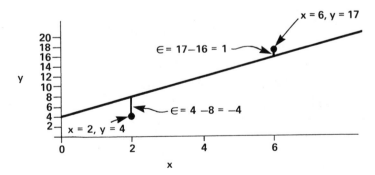

**FIGURE 9.4** Graph of the regression equation $y = 4 + 2x$ and residuals for the points $(x = 2, y = 4)$, and $(x = 6, y = 17)$.

makes the sum of the squared vertical deviations from the fitted line as small as possible). These are called **least-squares** estimates. Thus, suppose we have a sample of pairs and let us denote a particular pair (the i-th pair) $x_i$, $y_i$, so that

$$y_i = \beta_0 + \beta_1 x_i + \epsilon_i .$$

The i-th error is

$$\epsilon_i = y_i - \beta_0 - \beta_1 x_i,$$

as illustrated in Figure 9.4, and its square is

$$\epsilon_i^2 = (y_i - \beta_0 - \beta_1 x_i)^2.$$

Then the least-squares estimates $b_0$ and $b_1$ of $\beta_0$ and $\beta_1$ are those estimates of $\beta_0$ and $\beta_1$, respectively, that minimize the sum of these squared deviations over all the sample values. The slope ($\beta_1$ or its least-squares estimate $b_1$) is also called the **regression of y on x**, or the **regression coefficient of y on x.**

Notice that if the line provides a perfect fit to the data (i.e., all the points fall on the line), then $\epsilon_i = 0$ for all i. Moreover, the poorer the fit, the greater the magnitudes, either positive or negative, of the $\epsilon_i$. Now let us define the fitted line by $\hat{y}_i = b_0 + b_1 x_i$, and the estimated error, or **residual,** by

$$e_i = y_i - \hat{y}_i = y_i - b_0 - b_1 x_i.$$

Then a special property of the line fitted by least squares is that the sum of the $e_i$ over the whole sample is zero. If we sum the squared residuals, $e_i^2$, we obtain a quantity called the **error sum of squares,** or **residual sum of squares.**

If the line is horizontal ($b_1 = 0$), as in Figure 9.5, the residual sum of squares is equal to the sum of squares about the sample mean. If the line is neither horizontal nor vertical, we have a situation such as that illustrated in Figure 9.6. The deviation of the i-th observation from the sample mean, $y_i - \bar{y}$, has been partitioned into two components: (1) a deviation from the regression line, $y_i - \hat{y}_i = e_i$, the estimated error or residual; and (2) a deviation of the regression line from the mean, $\hat{y}_i - \bar{y}$, which we call the deviation due to regression (i.e., due to the straight-line model). If we square each of these three deviations ($y_i - \bar{y}$, $y_i - \hat{y}$, and $\hat{y}_i - \bar{y}$) and separately add them up

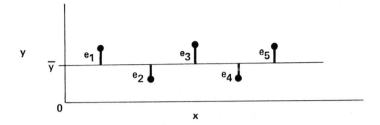

**FIGURE 9.5** Graph of the regression equation $\hat{y}_i = b_0 + b_1 x_i$ for the special case in which $b_1 = 0$ and $b_0 = y$, with five residuals depicted.

over all the sample values, we obtain three sums of squares which satisfy the following relationship

Total sum of squared deviations from the mean
= sum of squared deviations from the regression model + sum of squared deviations due to, or "explained by," the regression model.

We often abbreviate this relationship by writing

$$SS_T = SS_E + SS_R$$

where

$SS_T$ denotes the total sum of squares about the mean; $SS_E$ denotes the error, or residual, sum of squares; and $SS_R$ denotes the sum of squares due to regression.

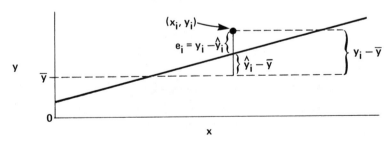

**FIGURE 9.6** Graph of the regression equation $\hat{y}_i = b_0 + b_1 x_i$, showing how the difference between each $y_i$ and the mean $y$ can be decomposed into a deviation from the line $(y_i - \hat{y}_i)$ and a deviation of the line from the mean $(\hat{y}_i - y)$.

These three sums of squares have $n - 1$, $n - 2$, and $1$ d.f., respectively. If we divide the last two sums of squares by their respective degrees of freedom, we obtain quantities called **mean squares:** the error, or residual, mean square and the mean square due to regression. These mean squares are used to test for the significance of the regression, which in the case of a straight-line model is the same as testing whether the slope of the straight line is significantly different from zero. In the example discussed above, we may wish to test whether the metabolite concentration in the reaction mixture is in fact changing, or whether it is the same at all the different points in time. Thus we would test the null hypothesis $H_0$: $\beta_1 = 0$. Denote the error mean square $MS_E$ and the mean square due to the straight line regression model $MS_R$. Then, under $H_0$ and certain conditions that we shall specify, the ratio

$$F = \frac{MS_R}{MS_E}$$

follows the F distribution with $1$ and $n - 2$ d.f. (Note, as is always true of an F-statistic, the first number of d.f. corresponds to the numerator, here $MS_R$, and the second to the denominator, here $MS_E$). As $\beta_1$ increases in magnitude, the numerator of the F-ratio will tend to have large values, and as $\beta_1$ approaches zero it will tend toward zero. Thus, large values of F indicate departure from $H_0$, whether because $\beta_1$ is greater or less than zero. Thus if the observed value of F is greater than the 95th percentile of the F distribution, we reject $H_0$ at the 5% significance level for a two-sided test. Otherwise, there is no significant linear relationship between Y and x.

The conditions necessary for this test to be valid are the following

1. For a fixed value of x, the corresponding Y must come from a normal distribution with mean $\beta_0 + \beta_1 x$.
2. The Y's must be independent.
3. The variance of Y must be the same at each value of x. This is called **homoscedasticity;** if the variance changes for different values of x, we have **heteroscedasticity.**

Furthermore, under these conditions, the least-squares estimates $b_0$ and $b_1$ are also the maximum likelihood estimates of $\beta_0$ and $\beta_1$.

**TABLE 9.1.** Summary Results for Testing the Hypothesis of Zero Slope (Linear Regression Analysis)

| Source of Variability in Y | d.f. | Sum of Squares | Mean Square | F-ratio |
|---|---|---|---|---|
| Regression | 1 | $SS_R$ | $MS_R$ | $MS_R/MS_E$ |
| Residual (error) | $n - 2$ | $SS_E$ | $MS_E$ | |
| Total | $n - 1$ | $SS_T$ | | |

The computation of $b_0$, $b_1$, and the mean squares, which nowadays are often automatically calculated on a computer or a pocket calculator, is detailed in Appendix 1. The statistical test is often summarized as shown in Table 9.1. Note that $SS_T/(n-1)$ is the usual estimator of the (total) variance of Y if we ignore the x-values. The estimator of the variance of Y about the regression line, however, is $MS_E$, called the **error mean square** or the **mean squared error**. It estimates the error, or residual, variance not explained by the model.

Now $b_1$ is an estimate of $\beta_1$ and represents a particular value of an estimator which has a standard error that we shall denote $s_{B_1}$. Some computer programs and pocket calculators give these quantities instead of (or in addition to) the quantities in Table 9.1. Then, under the same conditions, we can test $H_0$: $\beta_1 = 0$ by using the statistic

$$t = \frac{b_1}{s_{B_1}},$$

which under $H_0$ comes from Student's t distribution with $n - 2$ d.f. In fact, the square of this t is identical to the F-ratio defined earlier. So, in the case of simple linear regression, either an F-test or a t-test can be used to test the hypothesis that the slope of the regression line is equal to zero.

## THE STRAIGHT LINE RELATIONSHIP WHEN THERE IS INHERENT VARIABILITY

So far we have assumed that the only source of variability about the regression line is due to measurement error. But you will find that regression analysis is often used in the medical literature in situations in which it is known that the pairs of values x and y, even if measured without error, do not all lie on a line. The reason for this is not so much because such an analysis is appropriate, but rather because it is a relatively simple method of analysis, easily generalized to the case in which there are multiple independent variables (as we discuss later in this chapter), and readily available in many computer packages of statistical programs. For example, regression analysis might be used to study the relationship between triglyceride and cholesterol levels even though, however accurately we measure these variables, a large number of pairs of values will never fall on a straight line, but rather give rise to a scatter diagram similar to Figure 3.7. Regression analysis is not the appropriate statistical tool if, in this situation, we want to know how triglyceride levels and cholesterol levels are related in the population. It is, however, an appropriate tool if we wish to develop a prediction equation to estimate one from the other.

Let us call triglyceride level x and cholesterol level Y. Using the data illustrated in Figure 3.7, it can be calculated that the estimated regression equation of Y on x is

$$\hat{y} = 162.277 + 0.217x$$

and the residual variance is estimated to be 776 $(\text{mg}/100 \text{ ml})^2$. This variance includes both measurement error and natural variability, so it is better to call it "residual" variance rather than "error" variance. Thus, for a population of persons who all have a measured triglyceride level equal to x, we estimate that the random variable Y, measured cholesterol level, has mean $162.277 + 0.217x$ mg/100 ml and standard deviation $\sqrt{776} = 27.857$ mg/100 ml. This is the way we use the results of regression analysis to predict the distribution of cholesterol level (Y) for any particular value of triglyceride level (x). But we must not use the same equation to predict triglyceride level from cholesterol level. If we solve the equation

$$\hat{y} = 162.277 + 0.217x$$

for x, we obtain

$$x = -747.820 + 4.608\hat{y}.$$

This is exactly the same line, the regression of Y on x, although expressed differently. It is not the regression of the random variable X on y. Such a regression can be estimated, and for these same data turns out to be

$$\hat{x} = -41.410 + 0.819y$$

with residual variance 2,930 $(mg/100 \text{ ml})^2$. This is the equation that should be used to predict triglyceride level from cholesterol level. Figure 9.7 is a repeat of Figure 3.7 with these two regression lines superimposed on the data. The regression of Y on x is obtained by minimizing the sum of the squared **vertical** deviations (deviations in Y, i.e., parallel to the y-axis) from the straight line, and can be used to predict the distribution of Y for a fixed value of x. The regression of X on y, on the other hand, is the converse situation: it is obtained by minimizing the sum of the squared **horizontal** deviations (deviations in the random variable X, i.e., parallel to the x-axis), and it is the appropriate regression to use if we wish to predict the distribution of X for a fixed (controlled) value of y. In this latter case X is the dependent variable and y the independent variable.

FIGURE 9.7 Scatterplot of serum cholesterol versus triglyceride levels of 30 medical students, with the two estimated regression lines.

It is clear that these are two different lines. Furthermore, the line that best describes the single underlying linear relationship between the two variables falls somewhere between these two lines. It is beyond the scope of this book to discuss the various methods available for finding such a line, but you should be aware that such methods exist, and that they depend on knowing the relative accuracy with which the two variables X and Y are measured.

## CORRELATION

In Chapter 3 we defined variance as a measure of dispersion. The definition applies to a single random variable. In this section we introduce a more general concept of variability called **covariance**. Let us suppose we have a situation in which two random variables are observed for each study unit in a sample, and we are interested in measuring the strength of the association between the two random variables in the population. For example, without trying to estimate the straight line relationship itself between cholesterol and triglyceride levels in male medical students, we might wish to estimate how closely the points in Figure 9.7 fit an underlying straight line. First we shall see how to estimate the covariance between the two variables.

Covariance is a measure of how two random variables vary together, either in a sample or in the population, when the values of the two random variables occur in pairs. To compute the covariance for a sample of values of two random variables, say X and Y, with sample means $\bar{x}$ and $\bar{y}$, respectively, the following steps are taken

1. For each pair of values $x_i$ and $y_i$, subtract $\bar{x}$ from $x_i$ and $\bar{y}$ from $y_i$ (i.e., compute the deviations $x_i - \bar{x}$ and $y_i - \bar{y}$).
2. Find the product of each pair of deviations [i.e., compute $(x_i - \bar{x})(y_i - \bar{y})$].
3. Sum these products over the whole sample.
4. Divide the sum of these products by one less than the number of pairs in the sample.

Suppose, for example, we wish to compute the sample covariance for X,Y from the following data:

| $i$ | $x$ | $y$ |
|---|---|---|
| 1 | 10 | 30 |
| 2 | 20 | 50 |
| 3 | 30 | 70 |
| 4 | 40 | 90 |
| 5 | 50 | 110 |

Note that the sample means are $\bar{x} = 30$ and $\bar{y} = 70$. Following the steps outlined above we

1. Subtract $\bar{x}$ from $x_i$ and $\bar{y}$ from $y_i$:

| $i$ | $x_i - \bar{x}$ | $y_i - \bar{y}$ |
|---|---|---|
| 1 | $10 - 30 = -20$ | $30 - 70 = -40$ |
| 2 | $20 - 30 = -10$ | $50 - 70 = -20$ |
| 3 | $30 - 30 = \ \ \ 0$ | $70 - 70 = \ \ \ 0$ |
| 4 | $40 - 30 = \ \ 10$ | $90 - 70 = \ \ 20$ |
| 5 | $50 - 30 = \ \ 20$ | $110 - 70 = \ \ 40$ |

2. Find the products:

| $i$ | $(x_i - \bar{x})(y_i - \bar{y})$ |
|---|---|
| 1 | $(-20)(-40) = 800$ |
| 2 | $(-10)(-20) = 200$ |
| 3 | $(0)(0) = \ \ \ 0$ |
| 4 | $(10)(20) = 200$ |
| 5 | $(20)(40) = 800$ |

3. Sum the products: $800 + 200 + 200 + 800 = 2000$.
4. Divide by one less than the number of pairs:

$$s_{XY} = \text{sample covariance} = \frac{2000}{5-1} = 500.$$

Note in steps 1 and 2 that if both members of a pair are below their respective means (as in the case of the first pair, $i = 1$), the contribution to the covariance is positive ($+800$ for the first pair). It is similarly positive when both members of the pair are above their respective means ($+200$ and $+800$ for $i = 4$ and 5, in the example). Thus, a positive covariance implies that X and Y tend to covary in such a manner that when one is either below or above its mean, so is the other. A negative covariance, on the other hand, would imply a tendency for one to be above its mean when the other is below its mean, and vice versa.

Now suppose X is measured in pounds and Y in inches. Then the covariance is in pound-inches, a mixture of units that is difficult to interpret. Recall that a variance is measured in squared units and we take the square root of the variance to get back to the original units. Obviously this does not work for the covariance. Instead, we divide the covariance by the product of the estimated standard deviation of X and the estimated standard deviation of Y, which we denote $s_X$ and $s_Y$, respectively. The result is a pure, dimensionless number (no units) that is commonly denoted r and called the **correlation coefficient**, or **Pearson's product-moment correlation coefficient**, that is,

$$r = \frac{s_{XY}}{s_X s_Y}$$

where

$s_{XY}$ is the sample covariance of X and Y; $s_X$ is the sample standard deviation of X; and $s_Y$ is the sample standard deviation of Y.

Thus, in the above example, $s_X = \sqrt{250}$ and $s_Y = \sqrt{1000}$, so

$$r = \frac{500}{\sqrt{250}\,\sqrt{1000}} = 1.$$

In this example the correlation coefficient is 1. A scatterplot of the data indicates that all the points $(x_i, y_i)$ lie on a straight line with positive slope, as illustrated in Figure 9.8(a).

If all the points lie on a straight line with negative slope, as in Figure 9.8(b), the correlation coefficient is $-1$. These are the most extreme values possible: a correlation can only take on values between $-$ and $+1$. Figures 9.8(a–h) illustrate a variety of possibilities, and it can be seen that the magnitude of the correlation measures how close the points are to a straight line. Remember that the correlation coefficient is a dimensionless number, and so does not depend on the units of measurement. In Figures 9.8(a–h) the scales have been chosen so that the numerical value of the sample variance of Y is about the same as that of X—you can see that in each figure the range of Y is the same as the range of X. Now look at Figures 9.8(i) and (j).

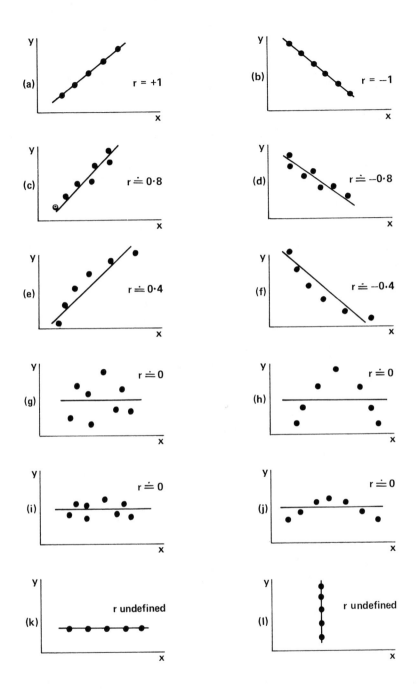

**FIGURE 9.8** Scatterplots illustrating how the correlation coefficient, r, is a measure of the linear association between two variables.

In each case the points appear to be close to a straight line, and you might therefore think that the correlation coefficient should be large in magnitude. If the scales are changed to make the range of Y the same as the range of X, however, Figure 9.8(i) becomes identical to Figure 9.8(g), and Figure 9.8(j) becomes identical to Figure 9.8(h). Once the scales have been adjusted, it becomes obvious that the correlation coefficient is near zero in each of these two situations. Of course, if all the points are on a horizontal line or on a vertical line, it is impossible to adjust the scales so that the range is numerically the same for both variables. In such situations, as illustrated in Figures 9.8(k) and (l), the correlation coefficient is undefined.

Note that the denominator in the correlation coefficient is the product of the sample standard deviations, which include both natural variability and measurement errors. Thus (unless the measurement errors in the two variables are themselves correlated), larger measurement errors automatically decrease the correlation coefficient. A small correlation between two variables can thus be due either to: (1) little linear association between the two variables; or (2) large errors in their measurement. A correlation close to $+1$ or $-1$, on the other hand, implies that the measurement errors must be small relative to the sample standard deviations, and that the data points all lie close to a straight line. In fact, there is a close connection between the correlation coefficient and the estimated slope of the regression line. The estimated slope of the regression of Y on x is $rs_Y/s_X$, and the estimated slope of the regression of X on y is $rs_X/s_Y$. The correlation coefficient is significantly different from zero if, and only if, the regression coefficients are significantly different from zero; and such a finding implies a dependency between the two variables. However, a correlation coefficient of zero does not imply two variables are independent [see Figure 9.8(h)], and as we have seen before, a dependency between two variables does not necessarily imply a causal relationship between them.

## SPEARMAN'S RANK CORRELATION

If we rank the x's from 1 to n (from largest to smallest, or vice versa), and we rank the y's from 1 to n in the same direction, and then compute the correlation coefficient between the pairs of ranks, we obtain the so-called **rank correlation coefficient** or **Spearman's rank correlation coefficient.** This correlation measures how closely the points can be fitted by a smooth, monotonic curve (i.e., a curve that is either always increasing or always decreasing). The rank correlations of the data in Figures 9.8(e) and (f) are $+1$, and $-1$, respectively. The curve that best fits the data in Figure 9.8(h), however, is not monotonic—it first increases and then decreases—and the rank correlation for these data is also approximately 0. Apart from being a measure of closeness to a monotonic curve, the rank correlation coefficient is less subject to influence by a few extreme values, and therefore sometimes gives a more reasonable index of association.

## MULTIPLE REGRESSION

We have seen how a straight-line regression model can be fitted to data so that the variable x "explains" part of the variability in a random variable Y. A natural question to ask is, if one variable can account for part of the variability in Y, can more of the variability be explained by further variables? This leads us to consider models such as:

$$Y = \beta_0 + \beta_1 x_1 + \beta_2 x_2 + \ldots + \beta_q x_q + \epsilon,$$

where $x_1, x_2, \ldots, x_q$ are q distinct independent variables. Just as before, we can partition the total sum of squared deviations from the mean of y as follows:

$$SS_T = SS_E + SS_R,$$

where $SS_E$ is the sum of squared deviations from the regression model and $SS_R$ is the sum of squared deviations due to, or "explained by" the regression model. But now $SS_E$ has $n - q - 1$ d.f. and $SS_R$ has q d.f. Following the same line of reasoning as used in the case of simple linear regression, we can compute the quantities indicated in Table 9.2. Thus, $F = MS_R/MS_E$ with q and $n - q - 1$ d.f. provides a simultaneous test of the hypothesis that all the regression coefficients are equal to zero; i.e.,

$$H_0: \beta_1 = \beta_2 = \ldots = \beta_q = 0 .$$

**TABLE 9.2.** Summary Results for Multiple Regression

| Source of Variability in Y | d.f. | Sum of Squares | Mean Square | F-ratio |
|---|---|---|---|---|
| Regression | $q$ | $SS_R$ | $MS_R$ | $MS_R/MS_E$ |
| Residual (error) | $n-q-1$ | $SS_E$ | $MS_E$ | |
| Total | $n-1$ | $SS_T$ | | |

The sum of squares for regression can be further partitioned into q terms, each with 1 d.f., so that each coefficient can be tested separately. Thus, the results of a multiple regression analysis with three independent variables might look something like Table 9.3. The F-test provides an overall test of whether the coefficients $\beta_1$, $\beta_2$, and $\beta_3$ are simultaneously zero. The t-tests provide individual tests for each coefficient separately. These t-tests are to be interpreted only in light of the full model. Thus, if we drop $x_2$ from the model because the results suggest $\beta_2$ is not significantly different from zero, then we can fit a new model, say

$$Y = \beta_0' + \beta_1'x_1 + \beta_3'x_3 + \epsilon$$

to make inferences about the coefficients of $x_1$ and $x_3$, with $x_2$ removed from the model. Note in particular that $\beta_1$ in the full model is not equal to $\beta_1'$ in the reduced model. Similarly, $\beta_3 \neq \beta_3'$.

**TABLE 9.3.** Summary Results of Regression Analysis for the Model $Y = \beta_0 + \beta_1x_1 + \beta_2x_2 + \beta_3x_3 + \epsilon$

| Source of Variability in Y | d.f. | SS | MS | F |
|---|---|---|---|---|
| Regression model | 3 | $SS_R$ | $MS_R$ | $MS_R/MS_E$ |
| Error (residual) | $n-4$ | $SS_E$ | $MS_E$ | |
| Total | $n-1$ | | | |

| Parameter | Estimate | Standard Error of Estimate | t | p-Value |
|---|---|---|---|---|
| $\beta_0$ | $b_0$ | $s_{B_0}$ | $b_0/s_{B_0}$ | |
| $\beta_1$ | $b_1$ | $s_{B_1}$ | $b_1/s_{B_1}$ | |
| $\beta_2$ | $b_2$ | $s_{B_2}$ | $b_2/s_{B_2}$ | |
| $\beta_3$ | $b_3$ | $s_{B_3}$ | $b_3/s_{B_3}$ | |

A special case of multiple regression involves fitting polynomial models to data. In this situation we may have measured only one independent variable x, but powers of x are also included in the regression model as separate independent variables. For example, we may fit such models as

$$Y = \beta_0' + \beta_1'x + \beta_2'x^2 + \epsilon \text{ (a quadratic model, or parabola)}$$

and

$$Y = \beta_0 + \beta_1x + \beta_2x^2 + \beta_3x^3 + \epsilon \text{ (a cubic model)}.$$

To study the effect of each independent variable fully, it is necessary to perform a regression analysis for every possible combination of the independent variables. In the example of Table 9.3, this would entail conducting a regression analysis for each of the following models:

1. Y regressed on $x_1$
2. Y regressed on $x_2$.
3. Y regressed on $x_3$.
4. Y regressed on $x_1$ and $x_2$.
5. Y regressed on $x_1$ and $x_3$.
6. Y regressed on $x_2$ and $x_3$.
7. Y regressed on $x_1$, $x_2$, and $x_3$.

The larger the number of independent variables, the greater the number of possible combinations, and so it is often not feasible to perform all possible regression analyses. For this reason a stepwise approach is often used, though it may not lead to the best subset of variables. There are two types of **stepwise regression:** forward and backward.

1. **Forward stepwise regression** first puts into the model the single independent variable that explains most of the variability in Y, and then successively at each step inserts the variable that explains most of the remaining (residual) variability in Y. However, if at any step none of the remaining independent variables explains a significant additional amount of variability in Y, at a predetermined level of significance, the procedure is terminated.

2. **Backward stepwise regression** includes all the independent variables in the model to begin with, and then successively at each step the variable that explains the least amount of variability in Y (in the presence of the other independent variables) is dropped from the model. However, a variable is dropped only if, at a predetermined level of significance, its contribution to the variability of Y (in the presence of the other variables) is not significant.

Whatever method is used to select among a set of independent variables in order to arrive at the "best" subset to be included in a regression model, it must always be remembered that the final result is merely a **prediction** model, and not necessarily a model for the **causation** of variability in the dependent variable.

## MULTIPLE CORRELATION AND PARTIAL CORRELATION

Each sample observation $y_i$ of the dependent variable corresponds to a fitted, or predicted, value $\hat{y}_i$ from the regression equation. Let us consider the pairs $(y_i, \hat{y}_i)$ as a set of data, and compute the correlation coefficient for these data. The result, called the **multiple correlation coefficient,** is denoted R. It is a measure of the overall linear association between the dependent variable Y and the set of independent variables $x_1, x_2, \ldots, x_q$ in the regression equation. In the special case that $q = 1$ (i.e., if there is only one independent variable), the multiple correlation coefficient R is simply equal to r, the correlation coefficient between X and Y. In general, $R^2$ equals the ratio $SS_R/SS_T$, and so measures the proportion of the variability explained by the regression model. If we fit a model with three independent variables and find $R^2 = 0.46$, and then fit a model which includes an additional, fourth independent variable and find $R^2 = 0.72$, we would conclude that the fourth variable accounts for an additional 26% (0.72 − 0.46) of the variability in Y. The multiple correlation coefficient $R^2$ is often reported when regression analyses have been performed.

The **partial correlation coefficient** is a measure of the strength of linear association between two variables after controlling for one or more other variables. Suppose, for example, we are interested in the correlation between serum cholesterol and triglyceride values in a random sample of men aged 20 to 65. Now it is known that both cholesterol and triglyceride levels tend to rise with age, so the mere fact that the sample includes men from a wide age range would tend to cause the two variables to be correlated in the sample. If we wish to determine the correlation after controlling for age (i.e., that part of the correlation that is over and above the correlation induced by common age), we would calculate the partial correlation coefficient, "partialing out the effect of" the variable age. The square of the partial correlation between the cholesterol and triglyceride levels would then be the proportion of the variability in cholesterol level that is accounted for by the addition of triglyceride to a regression model that already includes age as an independent variable. Similarly, it would also equal the proportion of the variability in triglyceride level that is accounted for by the addition of cholesterol to a regression model that already includes age as an independent variable.

## REGRESSION TOWARD THE MEAN

We conclude this chapter with a concept that, although it includes in its name the word "regression," and is indeed related to the original idea behind regression, is distinct from most of the ideas we have presented in this chapter. Consider the highest three triglyceride values among those listed in Table 3.1 for 30 medical students. They are (in mg/100ml) 218, 225, and 287, with a mean of 243.3. Suppose we were to draw aliquots of blood from these three students on several subsequent days; should we expect the mean of the subsequent values to be 243.3? Alternatively, if we took later samples from those students with the three lowest values (45, 46, and 49), should we expect the average to remain the same (46.7)? The answer is that we should not: we should expect the mean of the highest three to become smaller, and the mean of the lowest three to become larger, on subsequent determinations. This phenomenon is called regression toward the mean, and occurs whenever we follow up a selected, as opposed to a complete, or random, sample. To understand why regression toward the mean occurs, think of each student's measured triglyceride value as being made up of two parts: a "true" value (i.e., the mean of many, many determinations made on that student) and a random deviation from that true value—this latter could be due to measurement error and/or inherent variability in triglyceride value from day-to-day. When we select the three students with the highest triglyceride values based on a single measurement on each student, we tend to pick three that happen to have their highest random deviations—so that the mean of these three measurements (243.3 in our example) is most probably an overestimate of the mean of the three students' "true" values. Subsequent measures on these three students are equally likely to have positive or negative random deviations, so the subsequent mean will be expected to be the mean of their "true" values, and therefore probably somewhat lower. Similarly, if we pick the lowest three students in a sample, these single measurements will usually be underestimates of their true values, because they were probably picked partly because they happened to have their lowest random deviations when they were selected. If we were to make subsequent observations on the whole sample of 30 students, however, or on a random sample of them, regression toward the mean would not be expected to occur.

## SUMMARY

1. In the equation of a straight line $y = \beta_0 + \beta_1 x$, $\beta_0$ is the y-intercept and $\beta_1$ is the slope.

2. In simple linear regression analysis it is assumed that one variable (Y), the **dependent variable,** is subject to random fluctuations, whereas the other variable (x), the **independent variable,** is under the investigator's control. Minimizing the sum of the squared deviations of n sample values of Y from a straight line leads to the least-squares estimates $b_0$ of $\beta_0$ and $b_1$ of $\beta_1$, and hence the prediction line $\hat{y} = b_0 + b_1 x$. The sample residuals about this line sum to zero.

3. The total sum of squared deviations from the sample mean $\bar{y}$ can be partitioned into two parts: that due to the regression model and that due to error, or the residual sum of squares. Dividing these by their respective degrees of freedom gives rise to mean squares.

4. Under certain conditions the estimates $b_0$ and $b_1$ are maximum likelihood estimates, and the ratio of the mean squares (that due to

regression divided by that due to residual) can be compared to the F-distribution with 1 and $n-2$ d.f. to test the hypothesis $\beta_1 = 0$. These conditions are: (1) for a fixed x, Y must be normally distributed with mean $\beta_0 + \beta_1 x$; (2) the Y's must be independent; and (3) there must be **homoscedasticity**—the variance of Y must be the same at each value of x.

5. It is possible to determine a standard error for $b_1$ and, under the same conditions, $b_1$ divided by its standard error comes from a t distribution with $n-1$ d.f. In this situation the F-test and the t-test are equivalent tests of the null hypothesis $\beta_1 = 0$.

6. The regression of Y on x can be used to predict the distribution of Y for a given value x, and the regression of X on y can be used to predict the distribution of X for a given value y. The line that best describes the underlying linear relationship between X and Y is somewhere between these two lines.

7. Covariance is a measure of how two random variables vary together. When divided by the product of the variables' standard deviations, the result is the (product-moment) correlation, a dimensionless number that measures the strength of the linear association between the two variables. If all the points lie on a straight line with positive slope, the correlation is $+1$; if they all lie on a line with negative slope, the correlation is $-1$. A non-zero correlation between two random variables does not necessarily imply a causal relationship between them. A correlation of 0 implies no straight line association—but there may nevertheless be a curvilinear association.

8. The rank correlation (computed from the ranks of the observations) measures how closely the data points fit a monotonic curve. The rank correlation is not greatly influenced by a few outlying values.

9. Multiple regression is used to obtain a prediction equation for a dependent variable Y from a set of independent variables $x_1, x_2 \ldots$ . The significance of the independent variables can be jointly tested by an F-ratio, or singly tested by t-statistics. A stepwise analysis is often used to obtain the "best subset" of the x-variables with which to predict the distribution of Y, but theoretically we can only be sure of reaching the best subset by examining all possible subsets. The prediction equation obtained need not reflect any causal relationship between the dependent and independent variables.

10. The square of the multiple correlation coefficient is a measure of the proportion of the variability in a dependent variable that can be accounted for by a set of independent variables. The partial correlation coefficient between two variables is a measure of their linear association after allowing for the variables that have been "partialed out."

11. Whenever we make subsequent measurements on study units that have been selected for follow-up because they were extreme with respect to the variable being measured, we can expect regression toward the mean (i.e., study units with high initial values will tend to have lower values later, and study units with low initial values will tend to have higher values later).

## FURTHER READING

Kleinbaum, D.G. and Kupper, L.L. (1978). Applied Regression and Other Multivariate Methods. Duxbury Press. (This book covers many aspects of regression analysis, including computational formulae. It requires only a limited mathematical background to read.)

## PROBLEMS

1. Which of the following plots might represent data for variables X and Y that have a correlation coefficient equal to 0.82?

 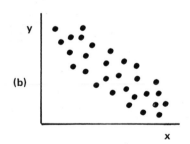

A. a
B. b
C. c
D. d
E. a and b

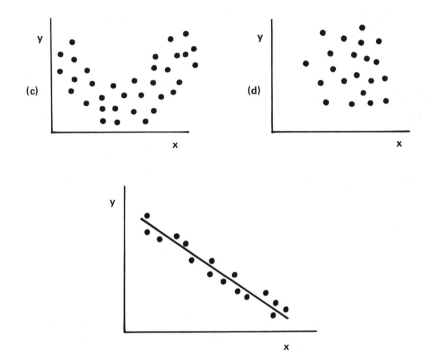

2. The correlation coefficient for the data in the above graph would be expected to have a value that is

A. a positive number of magnitude approximately equal to one
B. a negative number of magnitude approximately equal to one
C. a positive number of magnitude approximately equal to zero
D. a negative number of magnitude approximately equal to zero
E. none of the above

3. A group of men were examined during a routine screening for elevated blood pressure. Those men with the highest blood pressures—namely those with diastolic blood pressure higher than the 80th percentile for the group—were re-examined at a follow-up examination 2 weeks later. It was found that the mean for the re-examined men had decreased by 8 mmHg at the follow-up examination. The most likely explanation for most of the decrease is

A. the men were more relaxed for the second examination
B. some of the men became concerned and sought treatment for their blood pressure
C. observer bias

D. the observers were better trained for the second examination

E. regression toward the mean

4. Suppose two variables under study are temperature in degrees Fahrenheit (y) and temperature in degrees centigrade (x). The "regression line" for this situation is

$$y = \frac{9}{5}x + 32 .$$

Assuming there is no error in observing temperature, the correlation coefficient would be expected to be

A. $9/5$

B. $5/9$

C. $-1$

D. $+1$

E. $0$

5. A study was conducted to investigate the relationship between estriol level of pregnant women and subsequent height of their children at birth. A scatter diagram of the data suggested a linear relationship. Pearson's product-moment correlation coefficient was computed and found to be $r = 0.411$. The researcher decided to re-express height in inches rather than centimeters and then recompute r. The recalculation should yield the following value of r:

A.   0.000

B.  $-0.411$

C.   0.411

D.   0.500

E. cannot be determined from data available

6. An investigator finds the following results for a regression analysis following the model

$$Y = \beta_0 + \beta_1 x_1 + \beta_2 x_2 + \beta_3 x_3 + \epsilon$$

| Source of variability | d.f. | Sum of squares | Mean squares | F | p-value |
|---|---|---|---|---|---|
| Regression model | 3 | 120 | 40 | 1.43 | 0.25 |
| Error (residual) | 36 | 1008 | 28 | | |
| Total | 39 | 1128 | | | |

Assuming all assumptions that were made in the analysis are justified, an appropriate conclusion is
   A. the mean of the dependent variable Y is not significantly different from zero
   B. the intercept term $\beta_0$ is significantly different from zero
   C. the parameters of the model are significantly different from zero
   D. the independent variables $x_1$, $x_2$, and $x_3$ do not account for a significant proportion of the variability in Y
   E. none of the above

7. A forward stepwise regression analysis was performed according to the following models:

$$\text{Step 1: } Y = \alpha_0 + \alpha_1 x_1 + \epsilon$$
$$\text{Step 2: } Y = \beta_0 + \beta_1 x_1 + \beta_2 x_2 + \epsilon$$

The analysis was summarized as follows:

Step 1: $R^2 = 32.5\%$

| Source of variability | d.f. | Sum of squares | Mean squares | F | p-value |
|---|---|---|---|---|---|
| Regression model | 1 | 135 | 135 | 13.5 | 0.001 |
| Error (residual) | 28 | 280 | 10 | | |

Step 2: $R^2 = 47.5\%$

| Source of variability | d.f. | Sum of squares | Mean squares | F | p-value |
|---|---|---|---|---|---|
| Due to $x_1$ | 1 | 135 | 135 | | |
| Added to regression by $x_2$ | 1 | 62 | 62 | 7.68 | 0.01 |
| Error (residual) | 27 | 218 | 8.07 | | |
| Total | 29 | 415 | | | |

Assuming all assumptions made in the analysis are justified, an appropriate conclusion is
   A. $x_2$ accounts for a significant amount of the variability in Y over and above that accounted for by $x_1$

B.  neither $x_1$ nor $x_2$ accounts for a significant amount of the variability in Y
C.  the proportion of variability explained by the regression model containing both $x_1$ and $x_2$ is less than should be expected in a stepwise regression analysis
D.  the residual sum of squares is too large for meaningful interpretation of the regression analysis
E.  the F-ratio is too small for interpretation in step 2

8. A multiple regression analysis was performed assuming the model

$$Y = \beta_0 + \beta_1 x_1 + \beta_2 x_2 + \beta_3 x_3 + \epsilon.$$

The following results were obtained:

| Parameter | Estimate | Standard error of estimate | t-test | p-value |
|---|---|---|---|---|
| $\beta_0$ | 40 | 14.25 | 2.81 | 0.005 |
| $\beta_1$ | 5 | 2.43 | 2.06 | 0.025 |
| $\beta_2$ | 10 | 38.46 | 0.26 | 0.600 |
| $\beta_3$ | 7 | 2.51 | 2.79 | 0.005 |

Assuming all assumptions made in the analysis are justified, an appropriate conclusion is
  A.  none of the independent variables considered belong in the model
  B.  all of the independent variables considered belong in the model
  C.  $x_1$ and $x_2$ belong in the model but $x_3$ does not
  D.  $x_1$ and $x_3$ belong in the model but $x_2$ does not
  E.  $x_2$ and $x_3$ belong in the model but $x_1$ does not

9. An investigator studies 50 pairs of unlike-sex twins and reports that the regression of girl's birth weight (y) on boy's birth weight (x) is given by the following equation (all weights in grams):

$$y = 1221 + 0.403x$$

One can conclude from this that
  A.  the mean weight of twin brothers of girls who weigh 1,000 gm is predicted to be 1,624 gm

B.  the mean weight of twin sisters of boys who weigh 1,000 gm is predicted to be 1,624 gm
C.  the sample mean weight of the girls is 1,221 gm
D.  the sample mean weight of the boys is 1,221 gm
E.  the sample correlation between girl's weight and boy's weight is 0.403

10.  It is reported that both Pearson's product-moment correlation and Spearman's rank correlation between two variables are zero. This implies
A.  the two variables are independent
B.  the two variables tend to follow a monotonic curve
C.  there is no linear or monotonic association between the two variables
D.  all of the above
E.  none of the above

## CHAPTER 10

# Analysis of Variance and Linear Models

**Key Concepts:** analysis of variance: one-way or among and within groups, nested or hierarchical, two-way; sum of squares, mean square, expected mean square; sampling fraction, fixed effects, random effects; simple effects, main effects, interaction effects, additive effects; analysis of covariance, covariate; data transformation.

Methods for comparing the means of two groups were discussed in Chapter 7. We now discuss the comparison of means when there are three or more groups. As an illustration, suppose we wish to investigate the mean effects of three treatments A, B, and C. More specifically, we wish to use sample data to test the null hypothesis that the three population means are identical, i.e., we wish to test

$$H_0: \mu_A = \mu_B = \mu_C.$$

One approach to this problem is to perform all possible t-tests. In the present example, this involves three t-tests: one to test the hypothesis $\mu_A = \mu_B$, a second to test $\mu_A = \mu_C$, and a third to test $\mu_B = \mu_C$. There are a number of problems with this approach. In the first place, the tests are not independent; if $\mu_A = \mu_B$ and $\mu_A = \mu_C$, then it follows automatically that $\mu_B = \mu_C$. Thus, we can test any two of the three hypotheses and accomplish our goal. Which two of the three possible t-tests should we perform? In general, if there are $a$ means to be compared, we have $a - 1$ independent pairs and hence $a - 1$ degrees of freedom in choosing such pairwise comparisons.

One of the assumptions of the t-test is that the variances are equal in the two groups being compared. If we have three or more groups, we must assume all pairs of groups have equal variances in order to perform the pairwise t-tests in the usual manner. This is equivalent to assuming that all the groups have the same variance, however, and if this is so we can obtain a better estimate of the variance by pooling the information available in all the groups. Such a pooled estimate will have a greater number of degrees of freedom, and a test using this pooled estimate will have more power to detect differences among group means.

Even after eliminating the redundant pairwise comparisons and obtaining a pooled estimate of the variance, much effort will still be needed to compare all pairs of means with separate t-tests. It would be helpful if we had a single test to accomplish this goal. The F-test, which allows one to compare two variances, offers such a test. The strategy used is to compare the variability among the group means with the variability within the groups. If the variability among the sample group means is about what one might expect from the variability within the groups, we conclude that the group means are not significantly different. If, on the other hand, the variability among the sample group means is substantially larger than what we expect from the variability within the groups, we conclude that at least one of the pairwise comparisons is significant. Our criterion for evaluating the relative magnitude of these two sources of variability is the F-statistic. We compute a mean square deviation among group means and a pooled estimate of the mean square deviation within groups. Under the null hypothesis that the true population group means are equal, the ratio of these two mean squares—provided the observations are independent and come from normal distributions—follows an F distribution.

This approach to data analysis is an example of a general procedure, introduced by Sir Ronald Fisher, known as the analysis of variance. The analysis of variance (sometimes abbreviated as ANOVA) is essentially a procedure for partitioning the sum of squared deviations from the mean into components associated with recognized sources of variation. These components, which are sums of squares in their own right, are divided by their respective degrees of freedom to obtain mean squares. Ratios of mean squares are computed and compared with the F distribution to test hypotheses. Thus the analysis of variance is basically the same procedure as regression analysis. Whereas in regression analysis the independent variables are quantitative and usually continuous, however, in the analysis of variance the independent variables (such as "groups") are discrete categories.

In Chapter 2 we discussed a number of different experimental designs. For each design there is a corresponding analysis of variance. It is not possible in this brief chapter to cover the analysis corresponding to every design we have mentioned. Instead, we discuss the analysis of variance for a few special situations that illustrate the main principles involved.

## COMPLETELY RANDOMIZED DESIGN WITH A SINGLE CLASSIFICATION OF TREATMENT GROUPS

First we consider the situation in which we have $a$ treatment groups with the same number $n$ of observations in each group. There are a total of $na$ experimental units that have been randomly assigned, $n$ to each of the $a$ treatments. The results of any analysis of variance can be summarized in a table similar to the tables we have already seen for regression analysis. Table 10.1 shows the table appropriate for the situation we are considering. As before, we use the abbreviations SS for sum of squares and MS for mean square. Formulae for calculating the sums of squares are given in Appendix 1. Once the sums of squares have been calculated, the corresponding mean squares are obtained by dividing by the appropriate number of degrees of freedom. Thus, in Table 10.1, $MS_A = SS_A/(a - 1)$, and $MS_R = SS_R/a(n - 1)$. Finally, the F-ratio is computed. As indicated in Table 10.1, the F-ratio to test the significance of the group means is $MS_A/MS_R$. This particular analysis is sometimes called a **one-way**

**TABLE 10.1.** Outline of Analysis of Variance for Comparing $a$ Group Means With $n$ Observations per Group

| Source of Variability | Degrees of Freedom | Sum of Squares | Mean Squares | F |
|---|---|---|---|---|
| Among groups | $a - 1$ | $SS_A$ | $MS_A$ | $MS_A/MS_R$ |
| Within groups (residual) | $a(n - 1)$ | $SS_R$ | $MS_R$ | |
| Total | $an - 1$ | $SS_T$ | | |

**TABLE 10.2.** Summary Statistics for Diastolic Blood Pressure in mmHg After Four Weeks of Treatment

| | Treatment | | | |
|---|---|---|---|---|
| | A | B | C | D |
| Number of patients | 10 | 10 | 10 | 10 |
| Mean | 80 | 94 | 92 | 90 |
| Standard deviation | 10.5 | 9.5 | 9.7 | 10.2 |
| Standard error of the mean | 3.3 | 3.0 | 3.1 | 3.2 |

**analysis of variance,** or an **among and within groups analysis of variance.**

Suppose, for example, an experiment is conducted in which 40 patients are randomly assigned to four treatment groups, so that 10 patients receive each of the treatments A, B, C, and D, where the treatments are various drugs aimed at lowering blood pressure. Suppose further that after 4 weeks of treatment the results shown in Table 10.2 are obtained. Clearly, the mean diastolic blood pressure is lowest for treatment group A. But the question remains whether any of the differences are statistically significant. Since the four sample standard deviations are approximately equal, and diastolic blood pressure is approximately normally distributed, an F-test is appropriate for answering this question. The analysis of variance is given in Table 10.3. The F-ratio is 3.88, with 3 and 36 d.f. From Table 2, Appendix 2, we see that the 97.5th percentile of F with 3 and 36 d.f. lies between 3.54 and 3.45, and the 99th percentile lies between 4.51 and 4.31. Thus $0.01 < p < 0.025$ and we conclude that, at the 5% significance level, at least two of the means are different. It is clear from inspecting the mean blood pressures in Table 10.2 that the effect of treatment A is different from those of treatments B, C, and D, which are very much alike. Further statistical tests can be used (see

**TABLE 10.3.** Analysis of Variance Corresponding to the Summary Results in Table 10.2

| Sources of Variability | Degrees of Freedom | Sum of Squares | Mean Squares | F |
|---|---|---|---|---|
| Among drug groups | 3 | 1,160.00 | 386.67 | 3.88 |
| Within drug groups | 36 | 3,587.67 | 99.66 | |
| Total | 39 | 4,747.67 | | |

Appendix 1) to determine which sets of means are, and which are not, significantly different.

These concepts are sometimes made clearer by introducing a mathematical model for the data. Specifically, we introduce a **linear model** as follows: Let any observation be denoted $y_{ik}$, where i indicates the groups in which the observation belongs (i = 1, 2, . . . , a), and k denotes a specific observation in the i-th group (k = 1, 2, . . . , n). We take as our linear model

$$y_{ik} = \mu_i + \epsilon_{ik}$$

where $\mu_i$ is the mean of the i-th group and $\epsilon_{ik}$ is the random error, or residual, associated with the k-th observation in the i-th group. From this model we see that each observation is made up of two components, or effects: a mean effect, depending upon the group it belongs to, and a residual effect. Now consider the $\mu_i$ and $\epsilon_{ij}$ to be random variables. Let $\sigma_A^2$ be the variance of the $\mu_i$ (i.e., the variance among the population means), and let $\sigma_R^2$ be the variance of the $\epsilon_{ik}$ (i.e., the population residual variance). These are called **variance components.** Then if we take repeated samples, the expected (or mean) value of $MS_A$ is $\sigma_R^2 + n\sigma_A^2$; and the expected, or mean, value of $MS_R$ is $\sigma_R^2$. Note that under the null hypothesis the $\mu_i$ are all equal (and hence $\sigma_A^2 = 0$), and so these two mean squares then have the same expected value. It is necessary for two mean squares to have the same expected value if their ratio is to be distributed as F. As we shall see in later examples, this gives us a method of choosing the appropriate ratio of mean squares when we wish to test hypotheses in more complicated situations.

## DATA WITH MULTIPLE CLASSIFICATIONS

Let us now turn to the situation in which the data to be analyzed can be classified in more than one way. For instance, we may be able to classify the data both into the a groups $A_1, A_2, \ldots, A_a$ and also into the b groups $B_1, B_2, \ldots, B_b$. These are thus two "factors," A and B. The way to analyze the data then depends on the way the data are classified by these independent variables, or factors. Furthermore, when there are multiple classifications, the way we choose the treatments for investigation and the population to which we wish to make inferences also determine the appropriate test statistic for any particular hypothesis.

Let us assume that the $a$ treatments $A_1, A_2, \ldots, A_a$ are a subset of the total of $N_A$ possible treatments. Thus we have a sample of $a$ out of $N_A$ treatments, and we say the sampling fraction for treatments is $a/N_A$. The expected value of a mean square is often a function of the sampling fraction. If we do not wish to extrapolate beyond the treatments being investigated, $a = N_A$, and so the sampling fraction is 1. In this case we say the treatment effects are **fixed**. Suppose, for example, we have $a = 5$ nurses who are taking blood pressures and we wish to test the null hypothesis that there is no difference among the nurses with respect to the blood pressures they observe. If we are interested in making inferences only about these five particular nurses, then $N_A = 5$ and the sampling fraction is 1. We say that the "nurse effects" are **fixed**. On the other hand, suppose we consider these five nurses to be a random sample of all $N_A$ nurses in a hospital, and the question is whether there are significant differences among the nurses **in this hospital** with respect to the blood pressures they observe. Then $N_A$ is a larger number and the sampling fraction is less than 1. In this case we say the treatment effects (i.e., the "nurse effects") are **random**. If we wish to consider the five nurses to be a random sample of all nurses, so that we are testing whether there are differences among nurses in general with respect to blood pressures observed, then the sampling fraction is virtually zero.

The factors under study in an investigation may all be associated with fixed effects, they may all be associated with random effects, or some may be associated with fixed effects and others with random effects, so that we have a mixture of fixed and random effects. The models corresponding to these situations are called fixed models, random models, and mixed models, respectively.

## NESTED INDEPENDENT VARIABLES

Let us consider the situation in which the treatment categories of multiple factors follow a **nested**, or **hierarchical**, arrangement, such as might arise from a multi-stage cluster-sampling scheme. For exam-

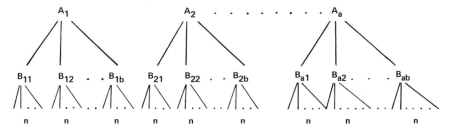

**FIGURE 10.1** Nested arrangement of treatment categories: $n$ replicate observations on each of $b$ aliquots from each of $a$ units of blood.

ple, suppose we have a sample of $a$ units of blood, which we denote $A_1, A_2, \ldots, A_a$. From each unit we take $b$ aliquots, and then on each aliquot we make $n$ replicate observations. This arrangement of the observations is illustrated in Figure 10.1, and the corresponding **nested, or hierarchical, analysis of variance** is outlined in Table 10.4. The units are the groups of type A and the aliquots are the groups of type B. Computational formulae for the sums of squares are given in Appendix 1 and the mean squares are obtained, as always, by dividing each sum of squares by its number of d.f. Note that within each group of observations of type A there are $b - 1$ d.f. among the groups of type B; hence the total number of d.f. among all the groups of type B, within the groups of type A (among aliquots within the $a$ units, in our example), is $a(b - 1)$. Similarly, there are $n - 1$ d.f. among the replicate observations within any group of type B, and hence, as there are altogether $ab$ groups of type B, there are $ab(n - 1)$ d.f. among replicates within groups of type B. Finally, note that the total

**TABLE 10.4.** Outline of Analysis of Variance for Two Nested Factors With $n$ Replicate Observations per Treatment

| Source of Variability | Degrees of Freedom | Sums of Squares | Mean Squares | F |
|---|---|---|---|---|
| Among groups of Type A | $a - 1$ | $SS_A$ | $MS_A$ | $F_A$ |
| Among groups of Type B within groups of Type A | $a(b - 1)$ | $SS_{B \mid A}$ | $MS_{B \mid A}$ | $MS_{B \mid A}/MS_R$ |
| Among replicates within groups of Type B (residual) | $ab(n - 1)$ | $SS_R$ | $MS_R$ | |
| Total | $abn - 1$ | $SS_T$ | | |

number of d.f. adds up to one less than the total number of observations, i.e.,

$$(a - 1) + a(b - 1) + ab(n - 1) = abn - 1.$$

This equality must hold for any analysis-of-variance table.

It can be seen in Table 10.4 that the appropriate F-statistic to test whether there are significant differences among the groups of type B (among aliquots within units, in our example) is $MS_{B|A}/MS_R$. The appropriate statistic ($F_A$) to test whether there are significant differences among the groups of type A, however, depends on the sampling fraction $b/N_B$. This can be seen by inspecting the expected mean squares in Table 10.5, which are functions of three variance components: $\sigma_A^2$, the variance among the group effects of type A; $\sigma_{B|A}^2$, the variance of the group effects of type B within the groups of type A; and $\sigma_R^2$, the variance of the residual effects within groups of type B. First, note that when $\sigma_{B|A}^2 = 0$, $MS_{B|A}$ and $MS_R$ have the same expected values, confirming that the F-statistic $MS_{B|A}/MS_R$ is appropriate to test for significant differences among the groups of type B, regardless of the sampling fraction $b/N_B$. To test for significant differences among the groups of type A, we need to divide $MS_A$ by a mean square that has the same expected value when $\sigma_A^2 = 0$ [i.e., by a mean square that has expected value $\sigma_R^2 + (1 - b/N_B)n\sigma_{B|A}^2$]. If $b = N_B$, then $1 - b/N_B = 0$ and $F_A = MS_A/MS_R$ is appropriate. If, on the other hand, $N_B$ is very large compared with $b$, then $b/N_B \doteq 0$ and $F_A = MS_A/MS_{B|A}$ is appropriate. Consider our example in which $b$ aliquots are taken from each unit of blood. A unit of blood comprises 500 ml. If we divide each unit into five 100 ml aliquots, so that $b = N_B = 5$, the aliquot effects are fixed and the divisor in $F_A$ should be $MS_R$. If, on the other hand, we take a sample of five 0.1 ml aliquots from a

**TABLE 10.5.** Expected Values of the Mean Squares for the Analysis of Variance in Table 10.4

| Mean Square | Expected Mean Square |
|:---:|:---:|
| $MS_A$ | $\sigma_R^2 + (1 - \dfrac{b}{N_B})n\sigma_{B|A}^2 + bn\sigma_A^2$ |
| $MS_{B|A}$ | $\sigma_R^2 + n\sigma_{B|A}^2$ |
| $MS_R$ | $\sigma_R^2$ |

total of 5,000 such aliquots, so that $b = 5$ and $N_B = 5,000$, then the aliquot effects are random; and because in this instance $b/N_B \doteq 0$, the divisor in $F_A$ should be $MS_{B|A}$. If $b/N_B$ is neither unity nor close to zero, then we must use an approximate test, but this is beyond the scope of this book.

This analysis of variance can be extended to any number of nested factors, and it is not necessary to have the same number of replicates within each treatment of a particular type, or the same number of treatment categories of a given type within each category of another type. There are computer programs that produce the analysis of variance table and calculate F-statistics. Many of these programs, however, use $MS_R$ as the divisor for all the F-tests, regardless of whether or not it is appropriate. In other words, many computer programs calculate F-statistics that are correct only if the effects of all the independent variables are fixed.

## CROSS-CLASSIFIED INDEPENDENT VARIABLES

In Chapter 2 we described the **factorial arrangement** of treatments, in which the treatments comprise all possible combinations of different levels of two or more factors. Thus if each factor is a drug and each level a different dose, then in a factorial arrangement the treatments comprise all possible combinations of dose levels of the drugs. Suppose, for example, we study two different dose levels of drug A in combination with three different dose levels of drug B, so that altogether there are six distinct treatments: $A_1B_1$, $A_1B_2$, $A_2B_1$, $A_2B_2$, $A_3B_1$, and $A_3B_2$. (In general we could have $a$ levels of factor A and $b$ levels of factor B, and so a total of $ab$ treatments). The resulting data are then **cross-classified**. The mean responses to the treatments can also be cross-classified, as the following two-way table shows:

|  |  | Factor B | | |
|---|---|---|---|---|
|  |  | $B_1$ | $B_2$ | $B_3$ |
| Factor A | $A_1$ | $\mu_{11}$ | $\mu_{12}$ | $\mu_{13}$ |
|  | $A_2$ | $\mu_{21}$ | $\mu_{22}$ | $\mu_{23}$ |

Thus, $\mu_{11}$ is the (population, or true) effect of the treatment $A_1B_1$, $\mu_{12}$ that of the treatment $A_1B_2$, and so forth. These means are called **simple effects**.

If in a study, $n$ sample units are randomly assigned to each treatment, we can describe the data by the following linear model

$$y_{ijk} = \mu_{ij} + \epsilon_{ijk},$$

where

$i = 1, 2$ (in general, $i = 1, 2, \ldots, a$); $j = 1, 2, 3$ (in general $j = 1, 2, \ldots, b$); and $k = 1, 2, \ldots, n$.

Thus, $\epsilon_{ijk}$ is a random amount by which the response of the k-th replicate differs from $\mu_{ij}$, the mean response (of all possible study units) to treatment $A_i B_j$. This model is similar to the model we introduced for analyzing data with a single classification factor, and in fact we can analyze the data as though there are just six different groups, testing the null hypothesis that the six group means (simple effects) are all equal.

When we have a factorial arrangement of treatments, however, it is often of interest to ask questions about each factor separately. We may ask, for example, whether there is any difference in mean response to $A_1$ and $A_2$, regardless of the level of B; or whether there are any significant differences among the mean responses to $B_1$, $B_2$, and $B_3$, regardless of the level of A. These mean responses are called **main effects.** Thus the main effect of treatment $A_1$ is the average of $\mu_{11}$, $\mu_{12}$, and $\mu_{13}$, and the main effect of $A_2$ is the average of $\mu_{21}$, $\mu_{22}$, and $\mu_{23}$. Similarly, the main effect of $B_1$ is the average of $\mu_{11}$ and $\mu_{21}$, and so on. These averages are usually taken to be unweighted averages [e.g., $(\mu_{11} + \mu_{12} + \mu_{13})/3$ for the main effect of $A_1$]. Sometimes, however, the main effects are defined as weighted averages, some simple effects being given more weight than others in the averaging process. Except in a special situation that will now be described, the difference between the main effects of $A_1$ and $A_2$, and the differences among the main effects of $B_1$, $B_2$, and $B_3$, depend on how the main effects are defined.

Suppose the differences between the simple effects of $A_1$ and $A_2$ are the same at all levels of B, i.e., suppose

$$\mu_{11} - \mu_{21} = \mu_{12} - \mu_{22} = \mu_{13} - \mu_{23}.$$

Then it does not matter how the simple effects are weighted in the definition of the main effects—the difference between the main effect of $A_1$ and the main effect of $A_2$ is always the same, and the difference between the main effects of any two particular levels of B is always the same. Equivalently, the same result holds (i.e., the differences between main effects do not depend on how the simple effects are weighted in the definition of the main effects) if the level of A has no effect on the difference between simple effects of $B_1$, $B_2$, and $B_3$, i.e., if

$$\mu_{11} - \mu_{12} = \mu_{21} - \mu_{22}$$

and

$$\mu_{12} - \mu_{13} = \mu_{21} - \mu_{23}.$$

In this situation we say there is **no interaction** between A and B, or that the effects of the different levels of A and B are **additive**. When the means are plotted on a graph and the simple effects making up each main effect are joined by straight lines, the lines are parallel, as in Figure 10.2. If the lines are not parallel, as in Figure 10.3, then an **interaction** is present and the different levels of A and B are not all additive. Thus, if $\mu_{11} - \mu_{21}$ is equal to $\mu_{12} - \mu_{22}$, which is the same as $\mu_{11} - \mu_{12}$ being equal to $\mu_{21} - \mu_{22}$, there is no interaction between the two levels of A and the first two levels of B; if these differences are not equal, then the difference between them, or

$$\mu_{11} - \mu_{21} - (\mu_{12} - \mu_{22})$$

[which is the same as $\mu_{11} - \mu_{12} - (\mu_{21} - \mu_{22})$] is an **interaction effect**.

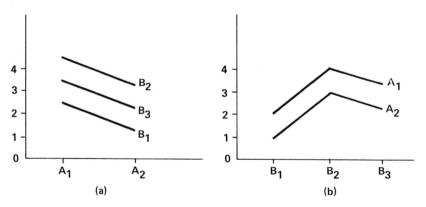

**FIGURE 10.2** Mean responses in a case in which there is no interaction: (a) for different levels of factor B plotted against the levels of factor A; (b) for different levels of factor A plotted against the levels of factor B.

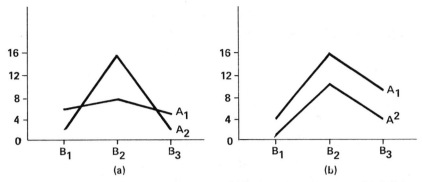

**FIGURE 10.3** Mean responses in a case in which there is interaction, the different levels of factor A being plotted against the levels of factor B. In case (a), no transformation can eliminate the interaction; in case (b) the square-root transformation eliminates the interaction.

Similarly

$$\mu_{12} - \mu_{13} - (\mu_{21} - \mu_{23})$$

is an interaction effect between the two levels of A and the second and third levels of B.

In practice, the presence or absence of interaction is obscured by chance variability in the data. We therefore test whether the interaction is significant (i.e., whether the observed departure of the sample means from additivity is too large to be explained by the chance variation within each treatment group). Similarly, we can test whether each of the main effects is significant (i.e., whether there are significant differences among the main effects of each factor). The analysis of variance table for $a$ levels of A and $b$ levels of B, with $n$ replicates of each treatment, is given in Table 10.6. This is often called a **two-**

**TABLE 10.6.** Outline of Analysis of Variance for a Factorial Arrangement of Treatments: Two Factors and $n$ Replicate Observations per Treatment

| Source of Variability | d.f. | Sum of Squares | Mean Squares | F |
|---|---|---|---|---|
| Main effects of A | $a - 1$ | $SS_A$ | $MS_A$ | $F_A$ |
| Main effects of B | $b - 1$ | $SS_B$ | $MS_B$ | $F_B$ |
| A × B interaction | $(a-1)(b-1)$ | $SS_{AB}$ | $MS_{AB}$ | $MS_{AB}/MS_R$ |
| Residual | $ab(n-1)$ | $SS_R$ | $MS_R$ | |
| Total | $nab - 1$ | $SS_T$ | | |

**way analysis of variance.** The F-statistic to test for the presence of interaction is $MS_{AB}/MS_R$, but the appropriate divisors in the statistics $F_A$ and $F_B$ to test for differences among the main effects depend on whether the effects are fixed or random.

Table 10.7 gives the expected values of the mean squares in terms of the following four variance components:

$\sigma_A^2$, the variance among the main effects of A.
$\sigma_B^2$, the variance among the main effects of B.
$\sigma_{AB}^2$, the variance among the interaction effects.
$\sigma_R^2$, the residual variance within the $ab$ treatments.

Inspection of this table reveals that if the main effects of a factor are fixed (so that the sampling fraction is 1), the appropriate divisor for the F-statistic to test for significant differences among the main effects of the other factor is $MS_R$. If the main effects of a factor are random and the sampling fraction is near 0, however, the appropriate denominator for the other factor is $MS_{AB}$.

Suppose, for example, we are interested in comparing two different types of surgery. Eligible patients are randomly assigned to each type of surgery and to each of four different surgeons. The two different types of surgery are fixed effects because they can hardly be considered a random sample of types of surgery: we chose these two particular types for study. We might wish to consider the four surgeons as a random sample, however, if we are interested in comparing the two types of surgery when performed by all surgeons, not just the particular four in our study. (Of course, the four surgeons were

**TABLE 10.7.** Expected Values of the Mean Squares for the Analysis of Variance in Table 10.6

| Mean Square | Expected Mean Square |
|:-----------:|:--------------------:|
| $MS_A$ | $\sigma_R^2 + (1 - \dfrac{b}{N_B})n\sigma_{AB}^2 + bn\sigma_A^2$ |
| $MS_B$ | $\sigma_R^2 + (1 - \dfrac{a}{N_A})n\sigma_{AB}^2 + an\sigma_B^2$ |
| $MS_{AB}$ | $\sigma_R^2 + n\sigma_{AB}^2$ |
| $MS_R$ | $\sigma_R^2$ |

most probably **not** a random sample of all surgeons; but if each is in a different location, and on the basis of several criteria they appear to be typical of the general population of surgeons, it could be of interest to analyze the results of the study as though the surgeons were a random sample.) In this situation, if we let the type of surgery be factor A, we have $a = N_A = 2$ with sampling fraction $a/N_A = 1$. Also, letting the different surgeons be factor B, we have $b = 4$ and $N_B$ large, so that the sampling fraction $b/N_B$ is near zero. Then, to test whether there is a significant difference between the two types of surgery, the appropriate divisor is $MS_{AB}$, whereas to test whether there are significant differences among surgeons the appropriate divisor is $MS_R$.

Notice that if we use $MS_{AB}$ as divisor when we should use $MS_R$, the F-statistic will tend on an average to be too small. The test will be **conservative,** tending too often not to reject the null hypothesis of no differences among the main effects. If, on the other hand, we use $MS_R$ as the divisor when we should use $MS_{AB}$, the test will tend on an average to be **anti-conservative**—the null hypothesis of no main effects will tend to be rejected too frequently. If there is no interaction ($\sigma_{AB}^2 = 0$), however, both $MS_{AB}$ and $MS_R$ have the same expected value and either can be used as the divisor. In this situation the best procedure is to pool the two mean squares to obtain an estimate of $\sigma_R^2$ with more d.f. The pooling is accomplished by adding together the corresponding sums of squares $SS_{AB}$ and $SS_R$, and then dividing by the sum of their d.f.:

$$(b - 1)(b - 1) + ab(n - 1).$$

This pooled mean square, with $(a - 1)(b - 1) + ab(n - 1)$ d.f., is then used as the divisor in F-tests for each of the main effects.

If there is only one study unit for each treatment (i.e., there is no replication and $n = 1$) there are no residual d.f. and $MS_R$ does not exist. Nevertheless, we can still test each of the main effects using $MS_{AB}$, with $(a - 1)(b - 1)$ d.f. as divisor. At worst, this test may be conservative (if $\sigma_{AB}^2 = 0$ and the main effects of one of the factors are random), and so not too powerful. If, however, one of the main effects is judged to be significant at a particular level on the basis of such a test, then we can be sure that there is indeed significance at that level.

If no interaction is present, it is easier to interpret a significant difference between two main effects. Sometimes it is possible to transform the measurements in such a manner that no interaction is present. For example, if we plot the square roots of the six means depicted in Figure 10.3(b), we obtain Figure 10.2(b). Thus there is no interaction among these means after transformation to the square-root scale. On the other hand, if there is a crossing of the lines as depicted in Figure 10.3(a), no transformation will eliminate the interaction. Biological systems are often governed by factors that act multiplicatively, rather than additively, in which case a logarithmic transformation will bring about additivity (i.e., remove interaction).

## ANALYSIS OF COVARIANCE

In both regression analysis and the analysis of variance we assume the data follow a linear model [i.e., that the dependent variable is a linear function of the independent variable(s) and a random error]. In regression analysis, the independent variables are quantitative and usually continuous, whereas in the analysis of variance the independent variables are always discrete. If our linear model contains both types of independent variable, the corresponding analysis is called an **analysis of covariance.**

As an example, suppose we wish to compare three different drug treatments for their effects on blood pressure, but, because we know blood pressure changes with age, we wish to include age in our linear model. Thus we have two independent variables: the quantitative variable age and the discrete variable drug treatment. The analysis would follow the same principles we have discussed for regression analysis and the analysis of variance, but would be called an analysis of covariance. The quantitative trait age is called a covariate, and the purpose of the analysis is to determine whether, after allowing for the effects of age, the treatment effects are significantly different. There are two reasons to allow for the effect of age. First, even though there is random assignment of patients to the different drug treatments, the three groups will probably still differ somewhat in their age distributions. An analysis of covariance compares the treatment effects "adjusted" to a common mean age for the three groups (i.e., as though the three groups had the same mean age). Second, even if the three groups have exactly the same mean age, allowing for age as a covariate identifies this extra source of variation and excludes it from the residual mean square. The residual mean square is thus smaller, and as a consequence, our tests of hypotheses are more powerful. Sources of variability that are randomly distributed among treatment groups do not affect the validity of significance tests, but do affect their power. An analysis that takes account of these sources, excluding from the residual mean square any variability due to them, is usually more powerful. **Concommitant variables,** which were discussed in Chapter 2, are for this reason often taken to be covariates in an analysis of covariance. The results of an analysis of covariance must be carefully scrutinized, however, because they can be misleading if the regression on a covariate (the slope of the straight line for predicting the dependent variable from that covariate) is not the same in all groups.

## ASSUMPTIONS ASSOCIATED WITH THE ANALYSIS OF VARIANCE

The F distribution is the theoretical model against which an F-statistic is evaluated. To be valid, hypothesis-testing procedures using the F distribution in the analysis of variance require the following assumptions:

1. The numerator and denominator of the F-ratio are independent.
2. The observations represent random samples from the populations being compared.
3. The observations are drawn from normally distributed populations.
4. The variances of all the populations are equal.

If the data fail to satisfy these assumptions, then the stated significance levels may be in error. For example, the F-tables may suggest that the p-value is 0.05 when the true p-value is 0.03 or 0.06. Whereas this small error may not be important, it is the uncertainty of its magnitude that is disturbing. Since it may be impossible to be certain that all the assumptions are satisfied exactly, F-tests in the analysis of variance are often viewed as approximate rather than exact.

The first two assumptions are satisfied if there is randomization. Random allocation of study units to the comparison groups virtually assures that the numerator and denominator of the F-ratio are independent, and random selection of the study units allows one to make valid inferences to populations.

Unless departure from normality is so extreme that it can be readily detected by visual inspection of the data, lack of normality has little effect on the test. The analysis of variance F-test to compare means is fairly robust against non-normality, whether in terms of skewness or kurtosis. This is in contrast to the F-test to compare two variances, discussed in Chapter 7, which is very sensitive to non-normality. In the case of marked skewness, it may be more appropriate to compare population medians rather than means. In such instances, it may be possible to transform the data (e.g., by taking logarithms or square roots of the data) to achieve a symmetric distribution of observations that more closely resembles a normal distribution. Analyzing the means on this transformed scale would be equivalent to analyzing the medians on the original scale.

Finally, the analysis of variance F-test is also robust to violation of the assumption of homogeneity of variances, provided the number of replicate observations is the same for each treatment. When the various treatment samples are unequal in size, however, large differences among the variances can have a marked effect on the F-test. A preliminary test can be performed to check the assumption of homoscedasticity. If significant heteroscedasticity is present, it may be possible to transform the data to achieve homogeneity among the variances.

## SUMMARY

1. The analysis of variance is a procedure for partitioning the sum of squared deviations from the mean into components associated with recognized sources of variation. The sums of squares are divided by their respective numbers of d.f. to obtain mean squares, and ratios of

mean squares are compared with the F distribution to test null hypotheses that sets of means are all equal, or that certain sources of variation are not significant.

2. Every analysis of variance is based on a particular linear model, and so the analysis of variance is basically the same procedure as regression analysis. Whereas in regression analysis the independent variables are quantitative and usually continuous, in the analysis of variance they are always discrete.

3. The appropriate analysis and F-tests depend on the experimental design. Simplest is a one-way, or among and within groups, analysis of variance. The null hypothesis is that the group means are all equal, and the F-statistic is the among-groups mean square divided by the within-groups, or residual, mean square.

4. When there are multiple classifications of the data, the appropriate divisor in the F-ratio may depend on the sampling fraction— the fraction of the total number of treatments of a particular type that are represented in the study. If the sampling fraction is 1, the treatment effects are fixed; if it is less than 1, they are random. For an F-statistic to follow the F distribution under $H_0$, it is necessary for both the numerator and denominator to have the same expected value under $H_0$.

5. In a nested, or hierarchical, analysis of variance, the residual mean square is the appropriate divisor to test for differences among groups at the lowest level within groups at higher levels. The appropriate mean square to test for differences among groups at higher levels depends on the sampling fraction(s) at lower levels.

6. In the analysis of cross-classified data, such as arise in a factorial arrangement of treatments, we define simple effects (cell means), main effects (averages of cell means that pertain to one level of a factor), and interaction effects (differences of simple effects that detect "non-parallelism"). If there are no interaction effects between the factors A and B, then the effects of the different levels of A and B are additive.

7. In a two-way analysis of variance, the residual mean square is appropriate to test for interaction effects. Either the residual or the interaction mean square is appropriate to test for one of the main effects, depending on whether the other main effects are fixed (sampling fraction = 1) or a random sample from a large population

(sampling fraction near 0). Use of the interaction mean square will always result in a test that is either valid or conservative, although perhaps not powerful. Sometimes interaction can be removed by a transformation of the data.

8. The analysis of covariance is a combination of regression analysis and the analysis of variance. It is often used to allow for concommitant variables when comparing several groups.

9. Analysis of variance F-tests assume: (1) the numerator and denominator are independent; (2) we have random samples; (3) we have normally distributed populations; and (4) the variances of all the populations are equal. Assumptions (1) and (2) are satisfied by appropriate randomization. The test is fairly robust against non-normality and also, provided the individual treatment samples are the same size, against heteroscedasticity.

## FURTHER READING

Dunn, O.J. and Clark, V.A. (1974). Applied Statistics: Analysis of Variance and Regression. John Wiley & Sons. (Chapters 5–7 give further computational details of the analysis of variance for designs we have discussed. The mathematical level of this text is fairly elementary.)

## PROBLEMS

1. Which of the following is **not** an assumption required in using the F distribution to test hypotheses in an analysis of variance?
   A. The numerator and denominator of the F-ratio are independent.
   B. The observations represent random samples from the populations being compared.
   C. The underlying linear model is a random model.
   D. The observations are drawn from normally distributed populations.
   E. The variances of all the populations are equal.

2. If the data in an analysis of variance fail to satisfy the required assumptions, then the F-tables may suggest that the p-value is 0.05 when in fact it is

A. $(0.05)^2$
B. exactly 0.10
C. less than 0.05
D. greater than 0.05
E. either less than or greater than 0.05

3. An investigator randomly assigned 10 patients to each of four different diets to study their effects on body weight. The resulting data were subjected to an analysis of variance. The F-test for the hypothesis that the mean response was the same for the four diet groups has degrees of freedom as follows:
A. numerator d.f. = 3, denominator d.f. = 10
B. numerator d.f. = 4, denominator d.f. = 9
C. numerator d.f. = 3, denominator d.f. = 36
D. numerator d.f. = 10, denominator d.f. = 40
E. numerator d.f. = 9, denominator d.f. = 36

4. A health-management corporation has a series of 12 clinics with five staff physicians at each clinic. The corporation wished to evaluate the differences among their clinics and physicians in managing blood pressure in hypertensive patients. Ten hypertensive patients were randomly assigned to each of two randomly selected physicians within each of four randomly selected clinics. An outline of the sources of variability, degrees of freedom, and expected mean squares in the analysis of variance resulting from the study is

| Sources of Variability | d.f. | Expected Mean Square |
|---|---|---|
| Clinics | 3 | $\sigma_R^2 + 6\sigma_{P|C}^2 + 20\sigma_C^2$ |
| Physicians within clinics | 4 | $\sigma_R^2 + 10\sigma_{P|C}^2$ |
| Patients within physicians within clinics | 72 | $\sigma_R^2$ |
| Total | 79 | |

The appropriate denominator for the F-test of the hypothesis that the clinic means are all the same is
A. the calculated mean square for clinics
B. the calculated mean square for physicians within clinics

C. the calculated mean square for patients within physicians within clinics

D. the calculated mean square for the total sample

E. none of the above

5. An experiment was conducted in which patients with chronic hypertension were administered one of two doses of drug A, $A_1$, or $A_2$, in combination with one of three doses of drug B, $B_1$, $B_2$, or $B_3$. In all there were six treatment groups. Forty-eight patients were randomly assigned to these groups, eight patients to a group. An analysis of variance of diastolic blood pressure after three weeks on treatment was carried out. The following table outlines the basis for an appropriate ANOVA.

| Source of Variability | d.f. | Expected Mean Square |
|---|---|---|
| Drug A | 1 | $\sigma_R^2 + 24\sigma_B^2$ |
| Drug B | 2 | $\sigma_R^2 + 16\sigma_B^2$ |
| Interaction | 2 | $\sigma_R^2 + 8\sigma_{AB}^2$ |
| Residual | 42 | $\sigma_R^2$ |
| Total | 47 | |

The appropriate denominator for the F-test of the hypothesis that the main effects of drug A are equal is

A. the calculated mean square for drug A

B. the calculated mean square for drug B

C. the calculated interaction mean square

D. the calculated residual mean square

E. none of the above

6. Consider the six treatment groups $A_1B_1$, $A_1B_2$, $A_1B_3$, $A_2B_1$, $A_2B_2$, and $A_2B_3$. Suppose the mean responses to these treatment combinations are as follows:

|  | $B_1$ | $B_2$ | $B_3$ |
|---|---|---|---|
| $A_1$ | 10 | 30 | 40 |
| $A_2$ | 12 | 33 | 41 |

The difference $33 - 30 = 3$ is

A. the difference between simple effects of A at $B_2$

B. the difference between main effects of A at $B_2$

C. an interaction of A at $B_2$
D. the simple effect of $B_2$
E. the main effect of $B_2$

7. Consider the six treatment groups $A_1B_1$, $A_1B_2$, $A_1B_3$, $A_2B_1$, $A_2B_2$, and $A_2B_3$. Suppose the mean responses to these treatment combinations are as follows:

|       | $B_1$ | $B_2$ | $B_3$ |
|-------|-------|-------|-------|
| $A_1$ | 21    | 31    | 34    |
| $A_2$ | 33    | 40    | 41    |

The average $(33 + 40 + 41)/3 = 38$ is called the
A. main effect of $A_1$
B. main effect of $A_2$
C. simple effect of $A_2$ at $B_1$, $B_2$, and $B_3$
D. interaction of A and B
E. error of A

8. Consider the four treatment groups $A_1B_1$, $A_1B_2$, $A_2B_1$, and $A_2B_2$. Suppose the mean responses to these treatment combinations are as follows:

|       | $B_1$ | $B_2$ |
|-------|-------|-------|
| $A_1$ | 50    | 52    |
| $A_2$ | 60    | 73    |

An estimate of the interaction effect between the factors A and B is
A. 2
B. 11
C. 13
D. 22
E. 26

9. A mixed model is one that has both
A. fixed effects and random effects
B. simple effects and main effects
C. interaction effects and main effects
D. interaction effects and fixed effects
E. interaction effects and random effects

10. In a study to investigate the effectiveness of an anticoagulant, 10 rats are randomly assigned to each of two groups. The rats in the first group receive an injection of the anticoagulant, while those in the second group receive a control saline injection. Samples of blood are taken from each rat before and after treatment and the coagulation time in minutes noted. The following method(s) might be appropriately used to test the hypothesis that the mean coagulation time is not affected by treatment:

(1) paired t test
(2) one-way analysis of variance of the pre-treatment values
(3) the correlation between pre-treatment and post-treatment values
(4) analysis of covariance

    A. 1, 2, and 3
    B. 1 and 3
    C. 2 and 4
    D. only 4
    E. 1, 2, 3, and 4

## CHAPTER 11

# Some Specialized Techniques

**Key Concepts:** multivariate analysis, discriminant analysis; logistic transformation or logit, logistic regression; survival time, singly and progressively censored data, survivorship function or curve, hazard function, life-table method, Kaplan-Meier method.

We have presented in the previous chapters basic concepts that should serve as building blocks for further study. These concepts have been illustrated by describing some of the common statistical methods found in the medical literature. It would, however, be impossible to cover in a single book all the statistical techniques that are referred to in medical journals. In this chapter we briefly familiarize you with some of the more specialized techniques of statistical analysis. Although the choice of which techniques to include and which to exclude is somewhat arbitrary, our aim has been to cover a few of the advanced methods of analysis that are more frequently encountered in medical research articles.

## MULTIVARIATE ANALYSIS

Interest often centers on the simultaneous analysis of several dependent variables rather than a single dependent variable. Let us suppose, for example, that an experiment is designed to study the effect of a treatment on the variables diastolic blood pressure, serum cholesterol, and body weight. In particular, let us suppose that the purpose of the study is to compare the means of these three dependent variables in a treated group to the corresponding means of these three variables in a control group. If one focused on a single variable (e.g. diastolic blood pressure), then one of the methods described earlier could be used to analyze the data. All the methods we have described so far are **univariate** methods, in that only one dependent variable is involved. (Sometimes multiple regression, in which there is more than one independent variable, is also called a multivariate method. This terminology, however, is incorrect if there is only one dependent variable.) If, on the other hand, all three variables are analyzed simultaneously as dependent variables, then the analysis is termed **multivariate.** For each of the univariate methods described earlier, there is an analogous multivariate method. For example, the multivariate analogue of Student's t-test for comparing the means of two groups is called Hotelling's $T^2$-test, named after the American statistician Harold Hotelling (1895–1973). Similarly, we can have multivariate regression analysis, multivariate analysis of variance, and multivariate analysis of covariance.

You may wonder what the advantage is to performing a multivariate analysis rather than performing a set of univariate analyses. Why not, in our example, simply perform three t-tests: one for diastolic blood pressure, one for serum cholesterol, and one for body weight? Multivariate analysis has basically three advantages. First, it helps overcome the problem that, as the number of statistical tests performed at a given significance level increases, so does the probability (under $H_0$) of finding at least one significant result. Suppose there are 20 dependent variables and we perform 20 t-tests at the 5% significance level (i.e., we decide beforehand to declare a significant finding if any one of the 20 p-values is less than 0.05). By chance alone, if there are no pairwise differences between the means, we should expect one of the tests to yield a significant result. If, however, we first perform a multivariate test, which takes account of the fact that 20 comparisons are being made, then we can appropriately control the overall probability of a Type I error to be, for example, 0.05. When we perform 20 t-tests each at the 0.05 significance level, the overall probability of a Type I error (i.e. the probability of rejecting the null hypothesis that all 20 pairs of means are equal when in fact they are equal) is much larger than 0.05.

A second advantage of multivariate tests is that they are more sensitive to detecting group differences that depend on certain relationships among the dependent variables. This can best be seen by considering a simple situation in which there are just two groups and two

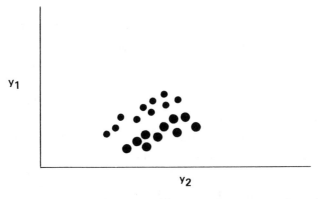

FIGURE 11.1 Scatterplot of the variables $y_1$ and $y_2$ measured on the ten study units in each of two groups, indicated • and ●.

dependent variables, $y_1$ and $y_2$, illustrated graphically in Figure 11.1. Here we have graphed, for two samples of 10 study units each, the values $y_1$ and $y_2$ observed on each study unit. It is clear from this scatter diagram that the two groups are completely separate, and a multivariate test of these data would show a highly significant difference between the two groups. But if we were to perform a t-test on the 20 values of $y_1$, or on the 20 values of $y_2$, neither result would be very significant because there is almost complete overlap between the two groups on each variable singly.

Lastly, there are certain methods of multivariate analysis that have no univariate analogue and can only be performed when there is more than one dependent variable. We now briefly discuss one such method.

## DISCRIMINANT ANALYSIS

Let us consider the situation in which it is unknown to which of two groups or populations a person belongs. For example, we may wish to know whether or not a woman is a carrier of the sex-linked hemophilia gene, and hence has a risk of bearing a son with hemophilia. Suppose we have laboratory data available on the woman and we wish to use the information to classify her. Using a procedure known as discriminant analysis, it is possible to determine a mathematical function for this purpose, from data on a set of previously classified women. Thus we would obtain two samples of women, one of women known to carry the gene (so-called "obligate carriers") and one of women known not to carry the gene. (If a woman has two hemophilic sons, for example, she is an obligate carrier; if she has no relatives with hemophilia, on the other hand, we can be virtually certain she does not carry the hemophilia gene.) A blood sample is taken from each woman and a set of relevant measurements, such as clotting-factor levels, determined. The result of a discriminant analysis applied to these data is a **discriminant function** that can be used to help classify a woman whose maternal uncle (but no other relative), for example, has hemophilia. In the case of hemophilia A, if we let

$$y_1 = \log \text{ (clotting factor XIII coagulant activity level)}$$

and $\quad y_2 = \log$ (clotting factor XIII-related antigen level),

the discriminant function derived from these two variables is (approximately) $3y_1 - 2y_2$. If this function is applied to the 20 points plotted in Figure 11.1, for example, the two groups are found to be distinct, with no overlapping of their ranges.

Discriminant analysis can also be used to classify individuals into one of several disease categories, based on vital signs and/or laboratory data. There will usually be errors associated with such classifications, and we try to develop discriminant functions that will correctly classify individuals with a high probability. Because we cannot know whether a particular individual is classified correctly, we often estimate the probability of an individual belonging to each population. The higher the probability associated with a person belonging to a particular disease category, the more confidence we have that we can correctly classify that person.

## LOGISTIC REGRESSION

In fitting a statistical model to a set of data, sometimes the dependent variable is dichotomous, whereas the independent variables are continuous, discrete, or both. In the simplest situations, we would have one dependent and one independent variable. For example, the dependent variable may be success or failure after a treatment, and we wish to model this response (the proportion of successes or failures) at selected doses of some treatment. In slightly more complex situations we may want to model the proportion of failures (e.g., the proportion of a population with disease) in terms of suspected risk factors such as age, weight, blood pressure, etc. In cases such as these, the cumulative distributions tend to be S-shaped, or tilted S-shaped, as in Figure 11.2. This characteristic shape arises because failures often occur infrequently at low levels of the independent variable(s), then there is a range in which the failure rate increases rapidly, and finally there is a range in which most of the failures have occurred and so additional failures occur less frequently again.

A family of mathematical equations that has a shape resembling that of Figure 11.2 is given by the equation

$$y = \frac{1}{1 + e^{-(\beta_0 + \beta_1 x)}}$$

or, equivalently,

$$y = \frac{e^{\beta_0 + \beta_1 x}}{1 + e^{\beta_0 + \beta_1 x}}$$

With further algebraic manipulation, this is the same as

$$\frac{y}{1 - y} = e^{\beta_0 + \beta_1 x},$$

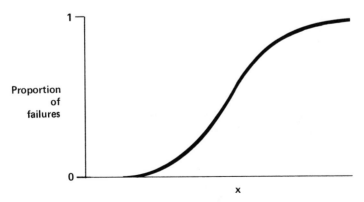

FIGURE 11.2 Example of a curve depicting the cumulative proportion of failures as a function of an independent variable, x.

or, taking natural logarithms of both sides of this equation,

$$\log_e \left[ \frac{y}{1-y} \right] = \beta_0 + \beta_1 x \ .$$

Thus, we have a transformation that converts the curve in Figure 11.2 into a straight line. This transformation is called the **logistic** or **logit transformation;** i.e., the logistic transformation, or logit of y is $\log_e [y/1-y]$. This is the basis of a **logistic regression** model, in which the logit of the dependent random variable Y (a proportion) is regressed on the independent variable x; i.e., the model is

$$\log_e \left[ \frac{Y}{1-Y} \right] = \beta_0 + \beta_1 x + \epsilon \ ,$$

where $\epsilon$ is a random error. There may also be several independent variables, in which case the model is of the form:

$$\log_e \left[ \frac{Y}{1-Y} \right] = \beta_0 + \beta_1 x_1 + \beta_2 x_2 + \ldots + \beta_p x_p + \epsilon \ .$$

A variety of computer programs are available for obtaining maximum likelihood estimates of the parameters $(\beta_0, \beta_1, \ldots \beta_p)$ of this model, and for testing hypotheses about them using the likelihood ratio criterion. The use of logistic regression models is becoming fairly common in the medical literature, especially for analyzing case-control studies. The corresponding estimated regression function (i.e., $b_0 + b_1 x_1 + b_2 x_2 + \ldots b_p x_p$) can also be used as a discriminant function to help classify future persons into one of the two classes (which in this instance we have called "successes" and "failures").

## ANALYSIS OF SURVIVAL TIMES

In some studies, especially clinical trials, the dependent variable of interest may be the amount of time from some initial observation until the occurrence of an event, such as recurrence of disease, death, or some other type of failure. This time from initial observation until failure is called the **survival time.** Statistical analysis of a group of survival times usually focuses on the probability of surviving a given length of time, or on the mean or median survival time.

A distinguishing feature of survival data is the fact that the distribution of survival times is often skewed and far from normal. Furthermore, the exact survival times of some of the study units may be unknown. For example, a group of patients may all enter a study at the same time, but some may not have "failed" by the end of the study, or they may be lost to follow-up at some point in the study. In such a case the survival times are said to be **censored**. If a study is conducted so that subjects are observed until a pre-specified proportion (e.g., 60%) have failed, or if all subjects are observed for a fixed period (e.g., five years), the resulting survival data are said to be **singly censored.** In most clinical studies, however, patients are recruited into the study over time, and each patient is observed for a different length of time. Then, if some of the patients have not failed by the end of the study, the resulting survival data are said to be **progressively censored.**

A distribution of survival times can be characterized by one of three functions: (1) the survivorship function, (2) the probability density function, and (3) the hazard function.

The **survivorship function,** S(t), is defined as the probability that a study unit survives longer than time t; i.e., if T is the random variable denoting survival time,

$$S(t) = P(T > t) .$$

S(t) is also known as the cumulative survival rate, and the graph of S(t) is called the **survival curve** (Figure 11.3). At any time t, S(t) gives the proportion still surviving at time t.

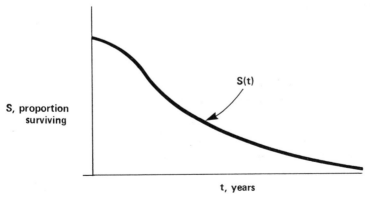

FIGURE 11.3 Example of a survival curve.

Recall that the cumulative distribution function of T is given by $F(t) = P(T \leq t)$. Hence it follows that

$$F(t) = 1 - S(t) .$$

The corresponding density function is the **probability density function f(t) of the survival time**. Areas under this curve represent the probability of failure in intervals of time.

The **hazard function** of survival time T is the density of failure at a particular time, given that there has been survival until that time. The hazard function is also known as the **instantaneous failure rate**, the **force of mortality**, or the **conditional failure rate**. It can be thought of as the "instantaneous" probability of failure at a particular time given there has been survival until that time. Because time is a continuous variable, however, it is a density rather than a probability. It is equal to $f(t)/S(t)$.

## ESTIMATING SURVIVAL CURVES

We shall describe two methods of estimating survival curves: (1) the life table method and (2) the Kaplan-Meier method.

In the life-table method, the survival times are first grouped into fixed intervals such as months or years. Let $n_i$ be the number of study units surviving at the beginning of the i-th interval, $d_i$ the number of failures in the i-th interval, and $c_i$ the number of censored survival times in the i-th interval. Then the probability that a study unit that has survived to the beginning of the i-th interval will survive through to the end of that interval is estimated as

$$s_i = \frac{n_i - d_i - c_i/2}{n_i - c_i/2} .$$

(It is assumed that the censored individuals leave randomly throughout the interval, so that on an average only half of them are present during the interval.) The overall probability of surviving until the end of the k-th interval is estimated as the product of the probabilities of surviving through each of the first k intervals, that is

$$s_1 s_2 \ldots s_k .$$

The Kaplan-Meier method of estimating survival curves uses the exact failure times rather than grouping them into intervals. Denote the ranked times of failure and censoring by $t_1 < t_2 < \ldots < t_i < \ldots < t_m$. Let $u_i$ be the number of units surviving at time $t_i$, and $f_i$ the number that fail at time $t_i$. A unit with survival time censored at time $t_i$ is assumed to survive up to and including time $t_i$. Then the probability that a study unit that has survived to time $t_{i-1}$ will survive to time $t_i$ is estimated as

$$q_i = \frac{u_i}{u_i + f_i}$$

As before, the overall probability of surviving to time $t_k$ is estimated as

$$q_1 q_2 \ldots q_k .$$

Expressions are available for the standard deviations of each of these estimates of the survival times.

## SUMMARY

1. Multivariate analysis is the simultaneous analysis of several dependent variables. It allows for appropriate control of the probability of a Type I error when many dependent variables are involved; and it can sometimes detect group differences that are not obvious when the variables are examined individually.

2. Every univariate method of analysis has a multivariate analogue. Some methods of multivariate analysis, such as discriminant analysis, have no univariate analogue. The purpose of discriminant analysis is to find a function of several variables that can help classify a study unit as coming from a particular population.

3. Logistic regression is used to model a proportion of a population (e.g., the proportion with a disease), or a probability, as a function of one or more independent variables. The logistic function transforms a tilted S-shaped curve into a straight line. The estimated regression function can also be used as a discriminant function (e.g., to help classify persons as having a disease or not).

4. Survival analysis is used when the dependent variable is a survival time (i.e., the time to a well-defined event, such as death). The distribution of survival times, which is usually skewed, may be characterized by a probability density function, a survivorship function (the complement of the cumulative distribution function), or a hazard function. The hazard function gives the density of failure at a particular time, given there has been survival up to that time.

5. Survival data are usually censored-either singly (if every unit has been observed for the same amount of time) or progressively (if the units have been observed for different lengths of time). Two methods of estimating survival curves (that allow for censoring) are the life-table method and the Kaplan-Meier method.

## PROBLEMS

1. An analysis is carried out to study the effects of three treatments on total serum cholesterol in patients with elevated cholesterol levels. The statistical model underlying the analysis included age of the patient as an independent variable. The resulting analysis is called
   A. analysis of covariance
   B. multivariate analysis
   C. discriminant analysis
   D. logistic regression analysis
   E. survival analysis

2. An experiment was conducted in which patients were randomly assigned to either an active treatment or a placebo. After three months of treatment, data were obtained for four variables: total serum cholesterol, serum triglyceride, systolic blood pressure, and diastolic blood pressure. A statistical analysis was carried out to test the null hypothesis that the treatment had no effect on any of the four variables. The resulting analysis is called
   A. univariate analysis
   B. discriminant analysis
   C. logistic regression analysis
   D. survival analysis
   E. multivariate analysis

3. An investigator studied two groups of patients: one group with confirmed coronary heart disease and a second in which overt coronary heart disease was not present. Total serum cholesterol, serum triglyceride, systolic blood pressure, and diastolic blood pressure were determined for each patient. The investigator wished to derive from these data a mathematical function that would help decide whether a patient with unknown coronary heart disease status, but on whom these four variables had been determined, has coronary heart disease. An appropriate statistical method for doing this is

A. univariate analysis
B. discriminant analysis
C. survival analysis
D. analysis of variance
E. analysis of covariance

4. Which of the following is an advantage of multivariate analysis?

A. The computations for it are simpler.
B. It requires fewer assumptions.
C. It allows for proper control of the Type I error when tests are performed on many dependent variables.
D. It avoids the requirement of randomization.
E. It always provides a more powerful test than a set of separate univariate analyses.

5. An investigator is studying the probability of disease in relation to several suspected risk-factor variables. A plot of the proportions with disease in various categories of each of the risk-factor variables indicates that each of the cumulative distributions is shaped like a tilted-S. This suggests that the investigator should consider a

A. univariate analysis
B. discriminant analysis
C. analysis of covariance
D. logistic regression analysis
E. none of the above

6. An investigator reported that the data from a study were analyzed using the Kaplan-Meier method. The investigator was most likely studying

A. multivariate data
B. survival data
C. discrete data
D. uncensored data
E. none of the above

7. An investigator wishes to estimate the instantaneous probability a patient will die, given that the patient has survived a given amount of time since an operation. In other words, the investigator is interested in estimating the following function of time since the operation:
A. probability density function
B. survivorship function
C. hazard function
D. cumulative distribution function
E. none of the above

8. A study of survival times of patients receiving coronary bypass operations is terminated while some of the patients are still surviving. For purposes of analysis, the survival times of these patients are said to be
A. discrete
B. multivariate
C. censored
D. terminated
E. none of the above

9. A researcher wishes to develop a statistical model to predict serum cholesterol levels based on a knowledge of five measures of dietary intake. The method for developing such a model can be described as
(1) multiple regression analysis
(2) discriminant analysis
(3) univariate analysis
(4) multivariate analysis
A. 1, 2, and 3
B. 1 and 3
C. 2 and 4
D. only 4
E. 1, 2, 3, and 4

10. An analysis is performed in which the proportion of persons with a disease in a group is divided by the proportion without the disease. A multiple regression analysis is carried out on the logarithm of the resulting ratio. This is an example of a general method known as

(1) correlation analysis
(2) multivariate analysis
(3) survival analysis
(4) logistic regression
    A.　1, 2, and 3
    B.　1 and 3
    C.　2 and 4
    D.　only 4
    E.　1, 2, 3, and 4

CHAPTER 12

# Guides to a Critical Evaluation of Published Reports

In reading a report published in the literature, one often begins by reading the abstract or summary. While this step is important in that it quickly indicates whether the article is really of interest, its role must be kept in perspective. You must not yield to the temptation of accepting conclusions from the summary without appraising the merit and validity of the study itself. You must read the article critically before accepting its conclusions as being relevant to the way you manage patients. Experience will improve your ability to evaluate research reported in the literature, but that ability will be best utilized if you approach your reading with a definite plan in mind. A wide variety of procedures are used to conduct, analyze, and report research findings, and so it is impossible to give a single set of hard and fast rules for evaluating all such reports. We have nevertheless compiled a few guidelines that you should find helpful to keep in mind as you read the literature.

## THE RESEARCH HYPOTHESIS

A first step in reviewing any article is to identify the research hypothesis. Why was this research performed? Does it have relevance to you? Is there any practical or scientific merit to it? If not, there is no need to read any further.

## VARIABLES STUDIED

When you have identified the research hypothesis, and before you read the report in detail, ask yourself what variables would shed light on the research hypothesis. Next, identify the variables included in the report. Which are the dependent variables? Which are the independent variables? Are these variables relevant to the research hypothesis? Compare your list of variables with those included in the report. Were important—possibly confounding—variables overlooked? Was adequate information collected on all relevant concomitant variables? Age, race, and gender are three important variables in many human studies. Ethnic origin, geographic location, and socioeconomic status are other variables that may be important.

The methods used to obtain the data are another important consideration. Were state-of-the-art clinical or laboratory techniques used? Do these methods produce precise measurements? The scale of measurement has an impact on the choice of statistical hypotheses to be tested and statistical methods for testing these hypotheses. Hence the scale of measurement used for recording each variable should be identified; some scales may be nominal, some ordinal, and some interval.

## THE STUDY DESIGN

It is imperative to identify the study design because the appropriate methods of analysis are determined by it. Is the study experimental or observational? What are the study units and how were they selected? Are they a random sample, or were they chosen merely because it was convenient to study these particular units? What was the target population? How does it compare with the study population? If, as is so often the case in experimental studies, a convenience (rather than a random) sample of study units was used, they should be described in sufficient detail for you to have some feel for how general any of the conclusions are. If the study units are patients at a particular hospital with some disease, for example, you must decide how representative these patients are. Examine the inclusion and exclusion criteria for entry into the study. Could the fact that these are patients at that particular hospital somehow make them unrepresentative? For example, does that hospital specialize in the more severe cases? Do geographic location or climatic conditions have an effect? In other words, can you reasonably use the results of this study to guide you in the treatment of **your** patients?

After the study units were selected, how was the study conducted? At what point was randomization used? Was an adequate control group included? Was double-blinding used to control observer bias?

## SAMPLE SIZE

The greater the number of study units in an investigation, the more confident we can be of the results. Findings based on one or two subjects cannot be expected to be typical of large populations. The sample size must be large enough for the results to be reliably generalized. Statistical tests used to declare results significant or not significant depend upon the sample size. The larger the sample size, the more powerful the test and the more sensitive the test to population differences. Hence, if the sample size is enormous, even trivial population differences could be declared significant. In a small sample, on the other hand, a large difference may be non-significant. Especially in the case of a study that reports no significant differences, it is necessary to determine how large a difference could have reasonably been detected by the sample size used. A good study that reports negative results will also quote the power of the statistical tests used.

## COMPLETENESS OF THE DATA

Clinical studies invariably suffer from the problem that some patients have missing data for one reason or another. If an unusually large number of study units, say more than 20%, have a significant amount of incomplete data, then the credibility of the results should be questioned. Be careful of reports in which a portion of the data have been discarded as outliers because of gross errors or other unusual circumstances. If an investigator discards those data that do not support the research hypothesis, the scientific objectivity is lost. A good study will include all the data available in the various analyses performed.

## APPROPRIATE DESCRIPTIVE STATISTICS

An overview of the findings of a study can often be gleaned by scanning the tables and graphs. Be watchful, however, that the tables or graphs are not misleading. Distinguish between the sample standard deviation and the standard error of an estimated parameter. If it is not reported, calculate the coefficient of variation for some of the important variables. If the relative variability is large (for example, if the coefficient of variation is greater than 30%), then important population differences may be obscured by the "noise" in the data. Be sure you understand what the numbers in the tables represent, and exactly what is graphed. It sometimes helps (especially in a poorly written report) to reconcile numbers in the tables and graphs with numbers in the text.

## APPROPRIATE STATISTICAL METHODS FOR INFERENCES

The names of statistical tests used in the analysis, such as Student's t-test, paired t-test, analysis of variance, multiple regression analysis, and so forth, should be stated in the methods section of the report. Be wary of reports that state p-values without indicating the specific method used to analyze the data. In each case, identify the specific null hypothesis that is being tested. Try to determine if the statistical methods used are appropriate for the study design, scale of measurement, etc. If the method of analysis is unfamiliar to you, consult a statistician. Each method requires certain assumptions of the data, such as independent samples, random samples, normal distributions or homgeneous variances. Gross violations of these assumptions may bias the analysis. The report of a careful analysis will justify the use of each statistical test used, should there be any doubt.

Remember that a p-value represents the chance that a difference in the sample data is the result of random variation, when in fact there is no difference in the populations from which the samples came. A p-value of 0.05 tells us there is a 5% chance that the observed difference (or a more extreme difference) could arise by chance if the null hypothesis is true. Suppose three independent statistical tests are carried out. If in each case the null hypothesis is true, the possibility that at least one of the tests results in significance at the 5% level is about 0.14. $[1 - (1 - 0.05)^3 \doteq 0.14]$ If the three tests are not independent, then the probability is somewhere between 0.05 and 0.14. If c comparisons are made, independent or not, the probability of obtaining, by chance, at least one significant at the $\alpha$ level is less than or equal to $1 - (1 - \alpha)^c$, which is approximately equal to $c\alpha$ for small values of

$\alpha$. In other words, if we want to correct a quoted p-value, p*, for the fact that c statistical tests have been performed, this can be done conservatively by multiplying it by c; we can be sure that the appropriate p-value is less than cp*. This is known as Bonferroni's method. Suppose, for example, that a sample of patients with a particular disease is compared with a sample of controls with respect to a panel of 50 HLA antigens, in order to determine if any of these antigens are associated with the disease. If any one of the tests results in a p-value of 0.001 (= 0.05 ÷ 50) or less, overall a significant result has been found at the 0.05 level.

## LOGIC OF THE CONCLUSIONS

Finally, there is no substitute for evaluating the logic of the conclusions. A report that concludes it is safer to drive at high speeds because relatively few deaths from automobile accidents occur at speeds in excess of 100 miles per hour is clearly absurd. The frequency of deaths occurring at speeds in excess of 100 miles per hour needs to be related to the number of cars driven over 100 miles per hour, and this compared with some similar fraction for cars driven at slower speeds. Equally absurd conclusions, however, can be found in many research reports because improper—or no—comparisons are made. Always be on your guard against this type of fallacy; it is more common than you may suspect.

## SUMMARY

1. Do not accept conclusions solely on the basis of the abstract or summary of a published report.

2. Once you have determined the research hypothesis, determine whether all important relevant variables were studied, and whether they were studied appropriately.

3. Identify the study design and the study population to determine the relevance of the results.

4. Take account of the sample size and the completeness of the data, especially if the study reports that no significant differences were found.

5. Be sure you understand what the numbers in the text and tables represent, and what is graphed.

6. Try to determine if the statistical methods used are appropriate. If multiple significance tests are performed, multiply each p-value by the number of tests performed to obtain an upper bound for the overall significance level.

7. Ask yourself if the conclusions make sense.

## FURTHER READING

Haines, S.J. (1981) Six Statistical Suggestions for Surgeons. Neurosurgery 9:414–417 (Some basic principles are given for interpreting statistical analyses in the medical literature. Catchy subheadings make this article enjoyable to read).

Glantz, S.A. (1980) Biostatistics: How to Detect, Correct and Prevent Errors in the Medical Literature. Circulation 61:1–7 (This article was referred to in Chapter 1; now is the time to read it.).

## PROBLEMS

1. Read articles in your own area of interest critically, following the guidelines of this chapter.

2. Read and critique the design and analysis aspects of the studies in the following two brief papers:

Gash, A. and Karliner, J.S. (1978) No Effect of Transcendental Meditation on Left Ventricular Function. Annals of Internal Medicine 88:215–216.

Michaels, R.M., Nashel, D.J., Leonard, A., Sliwinski, A.J., and Derbes, S.J. (1982) Weekly Intravenous Methotrexate in the Treatment of Rheumatoid Arthritis. Arthritis and Rheumatism 25:339–341.

# Epilogue

This book is different from many other books on biostatistics in that many detailed calculations and formulas have been excluded from the body of the text because they are not helpful in understanding the basic concepts involved. We have not stressed the details of calculating a regression line, for example, because the necessary calculations give little insight into the meaning of the estimates calculated. We have, however, included details of those calculations that will give the reader a better feel for the underlying concepts. Thus we have carefully explained how to calculate a sample standard deviation and a sample correlation coefficient, because these are cases in which going through the actual calculations will give the reader a better feel for what the estimates measure.

This book is also different from many introductory books in that it includes some topics that are usually covered only in mathematically more sophisticated texts. We have included, for example, the difference between estimates and estimators, maximum likelihood estimation, the likelihood ratio criterion, and random effects models in the analysis of variance. On the other hand, we have not described hypothesis testing as primarily a procedure for making a decision between two hypotheses, as is usually done, but rather as a means of quantifying disbelief in the null hypothesis by determining a p-value. Some statisticians may disagree with the choice of topics we have included as "essentials" of biostatistics, and with the way in which we have covered them. Now that you have read this book, however, you should have a basic understanding of most of the more common statistical methods used, and you should be in a good position to learn about many more that you may come across. Statistical methods are becoming more and more sophisticated every day. It would be impossible to cover in a book this size all the various statistical tests that occur in the literature of biomedical research. If a report you are reading refers to a particular statistical method you do not know about, consult a more advanced book or seek the help of a biostatistician who can explain the essentials of the method in simple terms.

If you need to use some of the simpler statistical methods, you may well be able to do so yourself with the help of this and/or other statistical textbooks. One need not always be a surgeon to remove a splinter. But, just as you would entrust major surgery to a specialist, so should you consult a competent biostatistician for major statistical work. Similarly, just as you can expect the surgeon to explain the surgery that will be performed, so can you expect the biostatistician to explain the statistical analysis that will be performed. You do not need the gory details, but you should understand the essentials. Of course there is nothing to stop you from learning more and perhaps becoming one day yourself a competent biostatistician, should you so desire. The purpose of our book, however, has been merely to help you understand the statistical content of biomedical research articles. If, in addition, this has the ultimate effect of upgrading the quality of statistical usage in your area of interest, we shall be more than repaid for our efforts.

# Review Problems

1. Statistics is used in medical research
   A. to make rational inferences
   B. to quantify uncertainty
   C. to summarize the results of experiments
   D. all of the above
   E. none of the above

2. The following are steps in the scientific method—the never-ending circle of refining our knowledge about the universe:
(a) test a hypothesis by experiment
(b) formulate hypothesis
(c) retain or reject hypothesis
(d) observe
The order of these steps should be
   A. dcba
   B. cbda
   C. adbc
   D. abcd
   E. dbac

3. The logic of inductive reasoning is an aspect of statistics that is used
   A. to prove mathematical theorems
   B. to make rational inferences about what happens in general
   C. to deduce specific results from general laws
   D. all of the above
   E. none of the above

4. You wish to conduct a clinical trial to compare two drugs in the treatment of a disease, and decide to enlist the help of a statistician. What is the best time to initiate contact with the statistician?

A. Before you have decided which two drug regimens to use, so that the statistician can help decide on what is medically appropriate.
B. After you have decided on what drug regimens to use, but before you decide on the sample of patients to receive them, so that the statistician can help design the trial.
C. After you have decided on the sample of patients, but before you actually conduct the trial, so that the statistician can help organize the data.
D. After the data have been collected, but before analyzing the data, so that the statistician can be responsible for the data analysis.
E. After the data are analyzed, but before the results are submitted for publication, so that the statistician can verify that an appropriate statistical analysis has been performed.

5. A physician develops a diagnostic test that is positive for 95% of the patients who have the disease and for 10% of the patients who do not have the disease. Of the patients tested, 20% have the disease. If the patient's test is positive, the probability a patient has the disease is approximately
A. 0.10
B. 0.30
C. 0.50
D. 0.70
E. 0.90

6. A 99% confidence interval implies that
A. the probability the given interval contains the true parameter is 0.99
B. the probability that 99% of the observations are in the interval is 0.99
C. on average, 99 out of 100 similarly constructed intervals would contain the true parameter
D. the probability the given interval does not contain the true parameter is 0.99
E. there is a 1% chance the hypothesis is false

7. An investigator collects diastolic blood pressure levels on a group of patients. He divides his scale of measurement into intervals of 5 mmHg (i.e., 70–74 mmHg, 75–79 mmHg, 80–84 mmHg, etc.). The investigator counts the number of patients with diastolic blood pressures in each interval. If the investigator were to plot the frequency of blood pressure levels in each interval, he would probably choose the following type of graph:

A. histogram
B. scatter diagram
C. regression line
D. bar graph
E. correlation coefficient

8. The figure below depicts a distribution that is

(1) platykurtic
(2) bimodal
(3) symmetric
(4) negatively skewed

A. 1, 2, and 3
B. 1 and 3
C. 2 and 4
D. 4 only
E. 1, 2, 3, and 4

9. An investigator states that the correlation between two variables is not statistically significant ($r = 0.07$). Which of the following conclusions is appropriate?

A. 0.07 should be quoted as the significance level.
B. 0.07 indicates a strong linear association between the two variables.
C. For every unit of change in one variable the other variable increases by 0.07 units.
D. High values of one variable tend to be associated with low values of the other variable.
E. Any association between the two variables does not appear to be linear.

10. Serum cholesterol levels in a group of young adults were found to be approximately normally distributed with mean 170 mg/100 ml

and standard deviation 8 mg/100 ml. Which of the following intervals includes approximately 95% of serum cholesterol levels in this group?

A. 160–180 mg/100 ml
B. 162–178 mg/100 ml
C. 150–190 mg/100 ml
D. 154–186 mg/100 ml
E. 140–200 mg/100 ml

11. The standard deviation of a population is 25. The standard error of the mean of a random sample of nine observations is approximately

A. 3
B. 8
C. 75
D. 225
E. 625

12. A difference is declared significant at the 5% significance level. This implies that
(1) the difference is significant at the 1% level
(2) the difference is significant at the 10% level
(3) there is a 5% probability that differences this large or larger could occur by chance
(4) there is 5% probability or less that differences this large or larger could occur by chance

A. 1, 2, and 3
B. 1 and 3
C. 2 and 4
D. only 4
E. 1, 2, 3, and 4

13. An estimator which on average gives the same value as the parameter being estimated is said to be

A. minimum variance
B. maximum likelihood
C. efficient
D. unbiased
E. symmetric

14. An investigator compares two treatments, A and B, and finds that the difference in responses for these treatments is not statistically significant (p = 0.25). This implies that
   A. the difference between the treatments is insufficient to say that the results could not have occurred by chance alone
   B. the probability the treatments are different in their effectiveness is 0.25
   C. the probability the treatments are equally effective is 0.25
   D. one of the treatments is 25% more effective than the other
   E. the difference in success rates using the two treatments is 25%

15. A patient checks her diastolic blood pressure at home and finds her average blood pressure for a two-week period to be 84 mmHg. Assume her blood pressure to be normally distributed with a standard deviation $\sigma = 5$ mmHg. A nurse checks her diastolic blood pressure in the office and finds a value of 110 mmHg. The office reading is apparently
   A. not atypical of her distribution of blood pressures
   B. consistent with normotensive diastolic blood pressure
   C. below the 95th percentile for this patient
   D. extremely high for this patient
   E. none of the above

16. "p-values" are reported often in the medical literature, yet their meaning is not always understood. The p-value is
   A. the power of the test
   B. the probability of getting a result as extreme or more extreme than the one observed if the null hypothesis is false
   C. the probability the null hypothesis is true
   D. the probability of making a Type II error
   E. the probability of getting a result as extreme or more extreme than the one observed, if the null hypothesis is true

17. An investigator found drug A to be more effective than drug B in reducing blood pressure (p = 0.02). A review of the literature revealed two other studies reported similar results, one finding p = 0.03 and the other p = 0.01. The probability that all three results

could have occurred by chance when in fact there was no difference
between the drugs is
A. 0.000006
B. 0.06
C. 0.94
D. 0.000094
E. 0.04

18. A Type II error
A. is often made when the p-value is small
B. is always made when the p-value is large
C. can only be made if the null hypothesis is true
D. can only be made if the null hypothesis is false
E. none of the above

19. An investigator wishes to test the hypothesis that the variance
of serum cholesterol levels in a group of school children is 20 (mg/100
ml)$^2$. Since the serum cholesterol levels in these children appear to be
normally distributed, the appropriate statistical distribution to use in
evaluating the test statistic is the
A. normal distribution
B. t-distribution
C. F-distribution
D. binomial distribution
E. chi-square distribution

20. Below are five tables giving the frequency of the presence of
symptoms among 200 patients with a particular disease and 200 patients without that disease. The rows and columns are: $D$ = disease
present; $\overline{D}$ = disease absent; $S$ = symptoms present; and $\overline{S}$ = symptoms absent. Which table presents results that would be expected to
yield the smallest chi-square value for testing the hypothesis that the
proportion of patients with symptoms present is the same whether or
not disease is present?

| A. | $\overline{D}$ | $D$ | Total | | B. | $D$ | $\overline{D}$ | Total |
|---|---|---|---|---|---|---|---|---|
| S | 100 | 100 | 200 | | S | 100 | 175 | 275 |
| $\overline{S}$ | 100 | 100 | 200 | | $\overline{S}$ | 100 | 25 | 125 |
| Total | 200 | 200 | 400 | | Total | 200 | 200 | 400 |

| C. | D | $\bar{D}$ | Total | | D. | D | $\bar{D}$ | Total |
|---|---|---|---|---|---|---|---|---|
| S | 175 | 100 | 275 | | S | 25 | 100 | 125 |
| $\bar{S}$ | 25 | 100 | 125 | | $\bar{S}$ | 175 | 100 | 275 |
| Total | 200 | 200 | 400 | | Total | 200 | 200 | 400 |

| E. | D | $\bar{D}$ | Total |
|---|---|---|---|
| S | 100 | 25 | 125 |
| $\bar{S}$ | 100 | 175 | 275 |
| Total | 200 | 200 | 400 |

21. The diastolic blood pressures of a random sample of 25 men are measured. The sample mean is found to be 85 mmHg, the standard deviation 5 mmHg. What are the approximate 95% confidence limits for the mean diastolic blood pressure of the population sampled?
   A. $85 \pm 1$ mmHg
   B. $85 \pm 2$ mmHg
   C. $85 \pm 5$ mmHg
   D. $85 \pm 10$ mmHg
   E. $85 \pm 15$ mmHg

22. Finding a p-value in significance testing begins with the basic assumption that
   A. well-trained observers record the data
   B. modern computing facilities are available
   C. the null hypothesis is true
   D. the sample size is at least 30
   E. the data are recorded on an interval scale

23. A researcher wishes to study the mean effects of four treatments, say A, B, C, and D, on triglyceride levels. He randomly assigns 40 patients to the four groups—10 patients to each group. He administers the treatments for two months and then measures the triglyceride level of each patient. His statistical hypothesis is that the mean triglyceride level is the same in each of the four groups. He knows from a plot of the data that, in these patients, log triglyceride levels are approximately normally distributed. The investigator should take logs of his data and use the following to test his hypothesis:

A. F-test
B. $\chi^2$-test
C. two-sample t-test
D. paired t-test
E. none of the above

24. Which of the following is a statistic?
A. The population standard deviation of height.
B. The mean of a binomial distribution.
C. The mean of a normal distribution.
D. The population variance of height.
E. The mean height of a sample of 10 men.

25. The power of a test is the probability of
A. accepting a true hypothesis
B. rejecting a true hypothesis
C. accepting a false hypothesis
D. rejecting a false hypothesis

26. Which of the following is based on statistical data?
(1) Smoking increases the risk of lung cancer.
(2) Wearing seat belts increases the chance of survival in automobile accidents.
(3) The chance a newborn baby is female is slightly less than 50%.
(4) The sun will rise tomorrow.
A. 1, 2, and 3
B. 1 and 3
C. 2 and 4
D. only 4
E. 1, 2, 3, and 4

27. Sometimes a one-tail test is used to determine significance level, and sometimes a two-tail test is used. The two-tail test is used for situations in which we know **a priori**
A. two samples are involved
B. both treatments will either increase or decrease the response
C. a paired t-test is appropriate
D. any true difference could be in either direction
E. two proportions are unequal

28. Assume the risk of developing stomach cancer during one's life is 1 in 50, whereas that of developing skin cancer is 1 in 25. Further assume that these events are independent. The risk of developing stomach cancer for a man who already has skin cancer is

A. 1 in 25
B. 1 in 50
C. 3 in 50
D. 1 in 1250
E. about 1 in 1.5 million

29. A distribution has a long tail to the right, so that it is not symmetric. We say the distribution is

A. abnormal
B. positively skewed
C. negatively skewed
D. bell shaped
E. bimodal

30. The probability an operation is a success is 3 in 4. The operation is performed on each of three patients. The probability at least one of the operations is a failure is

A. $1/64$
B. $63/64$
C. $27/64$
D. $37/64$
E. $48/64$

31. In a study of stomach cancer, patients who had the disease were matched with patients without cancer (controls) by age, sex, and social class. The frequency of alcohol consumption was then compared. What type of study was this?

A. Clinical trial.
B. Prospective.
C. Experimental.
D. Sample survey.
E. Retrospective.

32. A physician wishes to evaluate the effectiveness of three treatments, A, B, and C, with regard to time until relief of a headache.

He decides to study 60 patients with chronic headache problems by giving 20 patients A, 20 patients B, and 20 patients C, and observing them until they are relieved of their headaches. The physician randomly assigns A, B, or C to the 60 patients, using the restriction that 20 patients receive each treatment, but no other restriction. The design of this study is

A. double blind
B. change-over
C. completely randomized
D. latin square
E. randomized blocks

33. Advantages of using the double-blind procedure in clinical trials include
(1) to reduce observer variability in recording data
(2) to prevent biased observation of the outcome of a treatment
(3) to eliminate chances of a disproportionate number of poor-risk patients in one of the groups
(4) to reduce the probability of patients withdrawing from the study after discovering that they are in an undesired treatment group (e.g., a placebo group)

A. 1, 2, and 3
B. 1 and 3
C. 2 and 4
D. 4 only
E. 1, 2, 3, and 4

34. For a sample of n observations, we calculate the sample mean and then the deviation of each observation from the mean. These deviations are summed, and the result is divided by $n-1$; the result is

A. zero
B. the sample mean
C. the standard error of the mean
D. the sample standard deviation
E. the sample variance

35. Three new cases of a certain disease occurred during the month of July. If 500 persons were at risk during July, then the

A. prevalence was 3 per 1,000 persons
B. incidence was 3 per 1,000 persons
C. prevalence was 6 per 1,000 persons
D. incidence was 6 per 1,000 persons
E. odds ratio was 3:1

36. A group of patients was examined during a routine screening for elevated blood pressure. Twenty patients were told they had high blood pressure. A drug was prescribed for these patients and they were asked to return for re-examination one week later. At the second examination it was determined that the mean blood pressure for these patients was 10 mmHg lower than on the initial screening. It was claimed that the drug was responsible for the decrease. However, at least part, if not all, of the decrease would be expected to be due to a phenomenon known as
A. observer bias
B. double blinding
C. random allocation
D. false-negative testing
E. regression toward the mean

37. A physician conducted a sample survey of residents who graduated from a medical school during the last five years. The sample design was structured so that a random 10% of the graduates from each of the five classes were interviewed. The design can be described as a
A. simple random sample
B. stratified random sample
C. systematic random sample
D. two-stage random sample
E. random cluster sample

38. A control patient of the same age, race, and sex was found for each member of a group of 25 test patients. The control patients were given a standard drug and the test patients were given a new drug to determine whether the new drug increased the number of hours slept during one night of observation. An appropriate test for the hypothesis of interest is the

A. binomial test
B. paired t-test
C. two-sample t-test
D. F-test
E. chi-square test

39. An investigator studies the effect of a drug for lowering serum cholesterol levels. Patients are randomly assigned to either an active treatment group or a placebo group. The patients' cholesterol levels are then observed for one year while they take one of the treatments daily. This investigation can be described as a
(1) case-control study
(2) clinical trial
(3) retrospective study
(4) experimental study
    A. 1, 2, and 3
    B. 1 and 3
    C. 2 and 4
    D. only 4
    E. 1, 2, 3, and 4

40. The incidence of a certain disease among smokers was found to be 20 per 100,000 per year while among non-smokers it was found to be 5 per 100,000 per year. This implies that
    A. the risk of developing the disease is four times greater among smokers than among non-smokers
    B. smokers require four times as many hospital beds as non-smokers with the disease
    C. the prevalence of smoking was four times as great among diseased persons as compared to non-diseased persons
    D. the risk of developing the disease appears to be unrelated to smoking
    E. a clinical trial is needed to estimate the risk of developing the disease

41. In experimental studies, known sources of extraneous variability in the dependent variable are best controlled by using
    A. randomization
    B. completely randomized designs

C. randomized block designs

D. sample surveys

E. double-blind procedures

42. Investigator A claims his results are statistically significant at the 5% level. Investigator B argues that significance should be announced only if the results are statistically significant at the 1% level. From this we can conclude

    A. it will be more difficult for Investigator A to reject statistical null hypotheses if he always works at the 5% level (compared with Investigator B)

    B. it will be less difficult for Investigator A to reject statistical null hypotheses if he always works at the 5% level (compared with Investigator B)

    C. if Investigator A has significant results at the 5% level, they will also be significant at the 1% level

    D. if Investigator A has significant results at the 5% level they will never be significant at the 1% level

    E. none of the above

43. The duration of disease A is longer than that of disease B. They both have the same incidence. The prevalence of disease A would then be expected to be

    A. the same as that of disease B

    B. less than that of disease B

    C. greater than that of disease B

    D. less than the incidence of the disease

    E. greater than the incidence of the disease

44. Two laboratory methods for determining triglyceride levels are being compared. Method A has a coefficient of variation of 8%, while method B has a coefficient of variation of 3%. This implies that

    A. method B is less precise than method A

    B. the distribution of method B is less skewed than that of method A

    C. method B is less apt to produce false positives

    D. method B is easier to carry out than method A

    E. there is less bias in method A than in method B

45. Suppose the regression equation for predicting y from x is given by $\hat{y} = 10 + 3x$. Then
(1) the intercept is 10
(2) the correlation between x and y is positive
(3) the regression of y on x is 3
(4) the regression of x on y is 3
  A. 1, 2, and 3
  B. 1 and 3
  C. 2 and 4
  D. only 4
  E. 1, 2, 3, and 4

46. Which of the following is a parameter?
(1) The population standard deviation of height.
(2) The mean of a binomial distribution.
(3) The population variance of height.
(4) The mean of a normal distribution.
  A. 1, 2, and 3
  B. 1 and 3
  C. 2 and 4
  D. only 4
  E. 1, 2, 3, and 4

47. A researcher carries out an analysis of variance on a set of data, and finds a statistically significant interaction between two factors of interest. A similar analysis, performed after taking logarithms of the observed data and carried out on these transformed data, did not result in a statistically significant interaction. The analysis of the transformed data is the preferred analysis because then
  A. the effects of the two factors are additive
  B. the residual errors are uniformly distributed
  C. the residual errors are not independent
  D. the residual errors have variances proportional to the magnitude of the difference between the means for the individual factors
  E. none of the above

48. An analysis of variance was carried out to compare the mean diastolic blood pressure in four groups of patients. This resulted in the

finding that the variability among groups was not significantly greater than that within groups. A closer study of the data revealed, however, that the age distribution was not the same in the four groups. An analysis of covariance was therefore performed, with age as the covariate, and then statistically significant differences were found among the groups. The most likely explanation for this disparity in the results of the two analyses is

A. the residual errors are not independent
B. nonparametric tests are more powerful than their parametric counterparts
C. more precise comparisons among the groups are possible when the obscuring effect of age is accounted for
D. p-values are easier to obtain using the analysis of covariance procedure
E. none of the above

49. The main purpose of discriminant analysis is
A. to adjust for concomitant variables
B. to remove the obscuring effects of interaction terms
C. to explain linear relationships with covariates
D. to compare categorical data
E. to classify individuals into categories

50. The estimated probabilities of survival for the first, second, and third years after surgery in a group of experimental patients were found to be 0.80, 0.67, and 0.52, respectively, while those for a control group were found to be 0.76, 0.59, and 0.37. The difference between the two survival profiles was found to be statistically significant. Based on these data, the most appropriate conclusion is

A. the experimental group appears to have a better survival profile
B. the control group appears to have a better survival profile
C. the survival profiles are about the same
D. a regression analysis is needed to compare the two groups
E. incidence rates are required to compare the two groups

# Answers

## ANSWERS TO END-OF-CHAPTER PROBLEMS

*Chapter 2*

1—C. 2—E. 3—D. 4—B. 5—C. 6—C. 7—D. 8—B. 9—C. 10—D.

*Chapter 3*

1—B [The cumulative plot corresponds to the data shown in the histogram in Problem 2]. 2—A [The distribution is positively skewed: both median and mean are greater than the mode]. 3—B. 4—D. 5—B [Note that the range for Set 1 is 30, while for Set 2 it is 70]. 6—A. 7(1)—B. 7(2)—C. 8—B. 9—C. 10—A.

*Chapter 4*

1—D [0.007 + 0.020 − 0.0008]. 2—A [0.0008/0.020]. 3—E [0.0008/ 0.007]. 4—B. 5—D [0.3 × 0.3 × 0.3]. 6—E [1/3 + 2/3 × 2/3 or 1 − 2/3 × 1/3]. 7—C [0.7 × 0.8 or 1 − (0.3 + 0.7 × 0.2)]. 8—D [P(colorblind and male) = P(male)P(colorblind | male) = 0.5 × 0.05 = 0.025; P(colorblind and female) = P(female)P(colorblind | female) = 0.5 × 0.0025 = 0.00125; therefore P(male | colorblind) = 0.025/ (0.025 + 0.00125)]. 9—B [child receives 0 from father; the likelihood ratio is thus 1/0.67]. 10—D [joint probability for (2) is P(translocation)P(next child has Down's | translocation) = 0.05 × 0.16; for (3), (4): P(next child will have Down's) = P(translocation)P(next child will have Down's | translocation) + P(no translocation)P(next child will have Down's | no translocation) = 0.05 × 0.16 + 0.95 × 0.01 = 0.0175].

## Chapter 5

1—A. 2—E. 3—B. 4—A $[(1 - 0.02)^4]$. 5—C $[1 - (1 - 0.25)^3]$. 6—B [the number of possible mutant colonies on a plate is indefinitely large]. 7—D. 8—B $[15 \pm 2 \times 3]$. 9—B $[150 \pm 10]$. 10—D.

## Chapter 6

1—D [Note that the word "estimate" is commonly used in place of the more appropriate word "estimator"]. 2—D $[3 = s_{\bar{y}} = s_Y/\sqrt{n} = s_Y/\sqrt{9} = s_Y/3$; therefore $9 = s_Y$]. 3—C. 4—A [Compare t-values in the columns headed 99.5% and 97.5% in Table 1, Appendix 2]. 5—C. 6—B. 7—A. 8—A $[120 \pm 2(10/\sqrt{100})]$. 9—D. 10—B [For a large (n > 30) sample, the width of the 95% confidence interval is approximately $4s_{\bar{y}}$, i.e., $\bar{y} + 2s_{\bar{y}} - (\bar{y} - 2s_{\bar{y}})$, and so is proportional to $s_{\bar{y}}$. To halve the interval we need to halve $s_{\bar{y}}$. Since $s_{\bar{y}} = s_Y/\sqrt{n}$, n needs to be quadrupled: $s_Y/\sqrt{200} = \sigma_Y/\sqrt{4}\sqrt{50}$ is half of $\sigma_Y/\sqrt{50}$].

## Chapter 7

1—E. 2—B. 3—D. 4—B. 5—A. 6—C. 7—C. 8—A. 9—C. 10—D.

## Chapter 8

1—E. 2—D. 3—C $[(2 - 1)(4 - 1)]$. 4—A. 5—D. 6—D. 7—B. 8—D. 9—B. 10—B [If the proportions of exposed cases and controls are the same, the odds ratio is 1. The 95% confidence interval for the odds ratio does not include 1; this indicates that the proportions are significantly different at the 5% level (p < 0.05)].

## Chapter 9

1—A. 2—B. 3—E. 4—D. 5—C. 6—D. 7—A. 8—D. 9—B. 10—C.

*Chapter 10*

1—C. 2—E. 3—C. 4—E. 5—D. 6—A. 7—B. 8—B. 9—A. 10—D [using the pre-treatment value as a covariate].

*Chapter 11*

1—A. 2—E. 3—B. 4—C. 5—D. 6—B. 7—C. 8—C. 9—B. 10—D.

## ANSWERS TO REVIEW PROBLEMS

1—D. 2—E. 3—B. 4—B [Contact with a statistician should be initiated as early as possible, but it is not a statistician's role to help decide what is medically appropriate]. 5—D [Use Bayes' theorem: $\{(0.2)(0.95)/ (0.2)(0.95) + (0.8)(0.1)\}$]. 6—C. 7—A. 8—A. 9—E. 10—D. 11—B. 12—C. 13—D. 14—A. 15—D. 16—E. 17—A. 18—D. 19—E. 20—A. 21—B. 22—C. 23—A. 24—E. 25—D. 26—E. 27—D. 28—B. 29—B. 30—D. 31—E. 32—C. 33—C. 34—A. 35—D. 36—E. 37—B. 38—B. 39—C. 40—A. 41—C. 42—B. 43—C. 44—C. 45—A. 46—E. 47—A. 48—C. 49—E. 50—B.

# Additional Notes and Computational Formulae

## CHAPTER 3

1. The Greek capital sigma, $\Sigma$, is the mathematical sign for summation. If we have a sample of n observations, say $y_1, y_2, y_3, \ldots, y_n$, their sum is

$$y_1 + y_2 + y_3 + \ldots + y_n.$$

This also can be written $\sum_{i=1}^{n} y_i$, which is read as "the sum of the $y_i$, for i running from 1 to n." Thus the computational formula for the sample mean, which we shall denote $\bar{y}$, is

$$\bar{y} = \frac{1}{n} \sum_{i=1}^{n} y_i.$$

2. Using the same notation, the variance is

$$s^2 = \frac{1}{n-1} \sum_{i=1}^{n} (y_i - \bar{y})^2$$

$$= \frac{1}{n-1} \sum_{i=1}^{n} d_i^2, \text{ where } d_i = y_i - \bar{y}.$$

This is the best formula to use on a computer that can store all the $y_i$ in memory. If a calculator is being used, it is usually more convenient to use the mathematically equivalent formula

$$s^2 = \frac{1}{n-1} \left[ \sum_{i=1}^{n} y_i^2 - \frac{1}{n} \left( \sum_{i=1}^{n} y_i \right)^2 \right].$$

First $\sum_{i=1}^{n} y_i$ is calculated, squared, and divided by n — this gives the "correction factor" $1/n (\sum_{i=1}^{n} y_i)^2$. Then $\sum_{i=1}^{n} y_i^2$ is calculated and the "correction factor" subtracted from it. The result is divided by $n-1$.

3. The coefficients of skewness and kurtosis are respectively

$$\frac{n^{\frac{1}{2}} \sum_{i=1}^{n} (y_i - \bar{y})^3}{\left[ \sum_{i=1}^{n} (y_i - \bar{y})^2 \right]^{\frac{3}{2}}} \quad \text{and} \quad \frac{n \sum_{i=1}^{n} (y_i - \bar{y})^4}{\left[ \sum_{i=1}^{n} (y_i - \bar{y})^2 \right]^2}.$$

## CHAPTER 4

1. Bayes' theorem can be derived quite simply from the basic laws of probability. We have

$$P(C_j | B) = \frac{P(C_j \text{ and } B)}{P(B)}$$

and

$$P(B) = P(C_1 \text{ and } B, \text{ or } C_2 \text{ and } B, \ldots, \text{ or } C_k \text{ and } B)$$
$$= P(C_1 \text{ and } B) + P(C_2 \text{ and } B) + \ldots + P(C_k \text{ and } B),$$

because $C_1, C_2, \ldots, C_k$ are mutually exclusive. It follows immediately that

$$P(C_j | B) = \frac{P(C_j \text{ and } B)}{P(C_1 \text{ and } B) + P(C_2 \text{ and } B) + \ldots P(C_k \text{ and } B)}.$$

2. Data collected by a paternity testing laboratory can be used to estimate the proportion of past cases in which the alleged father was the true father, without any knowledge of whether in each case the alleged father was or was not the true father. Briefly, one method of doing this is as follows. From a large series of cases that have come to the laboratory we can calculate the proportion of cases that resulted in an exclusion, which we shall call $P(E)$. Using genetic theory and a knowledge of the gene frequencies in the population, we can calculate $P(E | C_2)$, the probability of exclusion given that a random man is the true father [see, for example, MacCluer and Schull (1963) Amer. J. Hum. Genet. 15:191]. If we assume only two mutually exclusive possibilities exist—$C_1$, the alleged father is the true father, or $C_2$, a random man is the true father—we must have

$$P(E) = P(C_1 \text{ and } E) + P(C_2 \text{ and } E)$$
$$= P(C_1)P(E | C_1) + P(C_2)P(E | C_2).$$

But $P(E \mid C_1)$—the probability of exclusion given that the alleged father is the true father—is zero (whenever the alleged father is the true father he will not be excluded by the blood testing). Therefore

$$P(E) = P(C_2)P(E \mid C_2),$$

and so we can calculate $P(C_2) = P(E)/P(E \mid C_2)$, and hence $P(C_1) = 1 - P(C_2)$. In situations in which this "prior probability of paternity" has been estimated it has been found to be typically between 0.65 and 0.9.

## CHAPTER 6

1. There is a subtle difference between the "probability" of a particular sample and its "likelihood." We use the word likelihood when we want to stress that we are interested in different values of the parameters, keeping the values of the random variables fixed at the values observed in a particular sample. We use the word probability, on the other hand, when we want to stress that we are interested in different values of the random variable, keeping the values of the parameters fixed at the population values. Thus the expression for $P(y$ male) on page 108 is both the probability and the likelihood of the sample, which term we use merely indicating whether we want to consider it as a mathematical function of $y$ or of $\pi$. Also, using the word "likelihood" avoids the awkward fact that when we are dealing with continuous random variables, any particular sample has (theoretically) zero probability of occurring. The **likelihood** of a particular $y$ in this case is the height of the probability density function at the point $y$, but, as explained in Chapter 5, that height is not a probability.

2. The asymptotic properties of unbiasedness, efficiency, and normality only hold in general for maximum likelihood estimators if the likelihood is made a maximum in the mathematical sense (i.e., when the estimates are substituted for the parameters, the likelihood has to be larger than when any other neighboring permissible values are substituted). A likelihood, just like a probability density function, can have one mode, several modes, or no modes; each modal value is a particular maximum likelihood estimate of the parameter. Thus, maximum likelihood estimates may not be unique, nor need they even exist for permissible values of the parameters. A variance, for

example, must be positive, and yet its maximum likelihood estimate may sometimes be negative. In a situation such as this the maximum likelihood estimate is sometimes said to be zero, if at zero the likelihood is largest for all permissible values of the parameter. In genetics, the recombination fraction cannot be greater than 0.5. For this reason the maximum likelihood estimate is often said to be 0.5 if the true maximum occurs at some value greater than 0.5. Since in these cases the likelihood is not, however, at a mathematical maximum (i.e., it is not at a mode), such estimators do not possess the same good properties, asymptotically, that true maximum likelihood estimators possess.

3. Even if a maximum likelihood estimate is obtained as a mathematical maximum of the likelihood, there are still certain cases in which, however large the sample, it is still not unbiased. This occurs if, as the sample becomes larger and larger, so does the number of unknown parameters. In such situations the problem is often overcome by maximizing a so-called "conditional likelihood." You are most likely to see this term in connection with matched-pair designs. You may also encounter the term "partial likelihood"; estimators that maximize this have the same usual asymptotic properties of maximum likelihood estimators.

4. To show the equivalence between

$$P(-2 \le \frac{\overline{Y} - \mu}{\sigma_{\overline{Y}}} \le 2) \doteq 0.95$$

and $P(\overline{Y} - 2\sigma_{\overline{Y}} \le \mu \le \overline{Y} + 2\sigma_{\overline{Y}}) \doteq 0.95,$

first note that within each probability statement there are two inequalities. We can manipulate these inequalities by the ordinary rules of algebra, without changing the probabilities, as follows (read the sign $<=>$ as "is equivalent to"):

$$-2 \le \frac{\overline{Y} - \mu}{\sigma_{\overline{Y}}} \le 2$$

$$<=> -2 \le \frac{\overline{Y} - \mu}{\sigma_{\overline{Y}}} \text{ and } \frac{\overline{Y} - \mu}{\sigma_{\overline{Y}}} \le 2$$

$$<=> -2\sigma_{\overline{Y}} \le \overline{Y} - \mu \text{ and } \overline{Y} - \mu \le 2\sigma_{\overline{Y}}$$

$$<=> \mu - 2\sigma_{\overline{Y}} \le \overline{Y} \text{ and } \overline{Y} \le \mu + 2\sigma_{\overline{Y}}$$

$$<=> \mu \le \overline{Y} + 2\sigma_{\overline{Y}} \text{ and } \overline{Y} - 2\sigma_{\overline{Y}} \le \mu$$

$$<=> \overline{Y} - 2\sigma_{\overline{Y}} \le \mu \le \overline{Y} + 2\sigma_{\overline{Y}}.$$

# CHAPTER 7

1. The null hypothesis must be very specific, since we need to know the distribution of the test criterion under it. If our research hypothesis is, for example, $\pi < 0.6$, we want to disprove the alternative $\pi \geq 0.6$. But, to be specific, we take as our null hypothesis $\pi = 0.6$. Clearly, if we reject the hypothesis $\pi = 0.6$ in favor of $\pi < 0.6$, we must reject with even greater conviction the possibility that $\pi > 0.6$. In other words, whenever an inequality is involved in what we try to disprove, we take as our specific null hypothesis the equality that is closest to the research hypothesis; for if we disprove that particular equality (e.g., $\pi = 0.6$), we shall have automatically disproved the whole inequality (e.g., $\pi \geq 0.6$).

2. The mean of the statistic T for the rank sum test can be derived as follows. First, note that the average of N numbers 1, 2, . . . , N is $(N + 1)/2$. Under the null hypothesis T is the sum of $n_1$ randomly picked numbers from the set of numbers 1, 2, . . . , $n_1 + n_2$. So, putting $N = n_1 + n_2$, their average is $(n_1 + n_2 + 1)/2$; therefore, the sum of $n_1$ random numbers from the set would be expected, on an average, to be $n_1(n_1 + n_2 + 1)/2$. If T is less than this we would suspect that the median of the first population is less than that of the second, and conversely if T is greater than this. It is not so simple to derive the standard deviation of T.

# CHAPTER 8

1. For any quantity $x^2$, we have

$$x^2 = (1 - \pi_1)x^2 + \pi_1 x^2.$$

Thus

$$\frac{(y_1 - n\pi_1)^2}{n\pi_1(1 - \pi_1)} = (1 - \pi_1)\frac{(y_1 - n\pi_1)^2}{n\pi_1(1 - \pi_1)} + \pi_1\frac{(y_1 - n\pi_1)^2}{n\pi_1(1 - \pi_1)}$$

$$= \frac{(y_1 - n\pi_1)^2}{n\pi_1} + \frac{(y_1 - n\pi_1)^2}{n(1 - \pi_1)}.$$

Now substitute $y_1 = n - y_2$ and $\pi_1 = 1 - \pi_2$ in the numerator of the second term; we obtain

$$\frac{(y_1 - n\pi_1)^2}{n\pi_1} + \frac{[(n - y_2) - n(1 - \pi_2)]^2}{n(1 - \pi_1)}$$

$$= \frac{(y_1 - n\pi_1)^2}{n\pi_1} + \frac{(-y_2 + n\pi_2)^2}{n\pi_2}$$

$$= \frac{(y_1 - n\pi_1)^2}{n\pi_1} + \frac{(y_2 - n\pi_2)^2}{n\pi_2}.$$

2. There is another (mathematically identical) formula for calculating the Pearson chi-square statistic from a $2 \times 2$ contingency table. Suppose we write the table as follows:

| a | b | (a + b) |
|---|---|---------|
| c | d | (c + d) |
| (a + c) | (b + d) | N = a + b + c + d . |

The formula is then

$$\frac{(ab - cd)^2 N}{(a + b)(c + d)(a + c)(b + d)}.$$

Sometimes this formula is modified to include a so-called "correction for continuity" that makes the resulting chi-square smaller. Provided each expected value is at least 5, however, there is no need for this modification.

3. Suppose we calculate the usual contingency table chi-square to test for independence between two dependent traits when we have matched pairs. As in Chapter 8, suppose each pair consists of a man and a woman matched for age, and the dependent variables are cholesterol level of the man and cholesterol level of the woman in each pair. Then the usual contingency table chi-square would be relevant for answering the question: is the cholesterol level of the woman in the pair independent of the cholesterol level of the man in the pair? (i.e., Does age, the matching variable, have a common effect on the cholesterol level of both men and women?) In other words, in this type of study the usual contingency table chi-square tests whether the matching was necessary—a significant result indicating that it was, a nonsignificant result indicating it was not.

4. The likelihood ratio criterion is a general criterion for testing one hypothesis against any other. Often we test a specific null hypoth-

esis under a general statistical model: the null hypothesis is a special case of the model. In the example in Chapter 8, the model for the two samples of data is that they come from two normal distributions, both of which have the same variance. Under this model, the means of the two distributions could be the same or different. We test the null hypothesis that they are the same. In order for $-2\log_e$ (likelihood ratio) to be asymptotically distributed as chi-square, it is necessary for the null hypothesis to be a special case of a general model. The numerator of the likelihood ratio is maximized under the null hypothesis, and the denominator is maximized under the model. The likelihood ratio then lies between 0 and 1, and $-2\log_e$ (the likelihood ratio) cannot be negative.

## CHAPTER 9

1. For a sample of n pairs, let $\bar{x}$ and $\bar{y}$ be the sample means. Define the quantities:

$$SS_x = \sum_{i=1}^{n} (x_i - \bar{x})^2 = \sum_{i=1}^{n} x_i^2 - \frac{1}{n}\left(\sum_{i=1}^{n} x_i\right)^2$$

$$SS_y = \sum_{i=1}^{n} (y_i - \bar{y})^2 = \sum_{i=1}^{n} y_i^2 - \frac{1}{n}\left(\sum_{i=1}^{n} y_i\right)^2$$

$$SS_{xy} = \sum_{i=1}^{n} (x_i - \bar{x})(y_i - \bar{y}) = \sum_{i=1}^{n} x_i y_i - \frac{1}{n}\left(\sum_{i=1}^{n} x_i\right)\left(\sum_{i=1}^{n} y_i\right).$$

Thus $SS_x$ is the total sum of the squared deviations of the $x_i$ from the mean $\bar{x}$, and $SS_y$ is the total sum of the squared deviations of the $y_i$ from the mean $\bar{y}$. $SS_{xy}$ is the analogous total sum of the cross-products; and the same considerations govern which formula to use for calculating this as govern the calculation of $SS_x$ and $SS_y$. (See note 2 for Chapter 4 in this Appendix.) Using these quantities, we successively calculate the estimated regression coefficient $b_1$ and intercept $b_0$, for the regression line $\hat{y}_i = b_0 + b_1 x_i$, as follows

$$b_1 = \frac{SS_{xy}}{SS_x}$$

$$b_0 = \bar{y} - \frac{SS_{xy}}{SS_x}\bar{x} = \bar{y} - b_1\bar{x}.$$

The sum of squares in Table 9.1 can then be calculated as

$$SS_R = \frac{SS_{xy}^2}{SS_x} = b_1 SS_{xy}$$

$$SS_E = SS_y - \frac{SS_{xy}^2}{SS_x} = SS_y - b_1 SS_{xy}$$

and, as explained, the mean squares are obtained by dividing the sums of the squares by their respective d.f.; i.e.,

$$MS_R = \frac{SS_R}{1} = SS_R$$

$$MS_E = \frac{SS_E}{n-1} \quad .$$

Note that the total sum of squares is

$$SS_T = SS_R + SS_E = SS_y,$$

(i.e., the total sum of squares of the dependent variable Y about its mean). The standard error of $b_1$ is $\sqrt{MS_E/SS_x}$.

2. The estimated regression line of **X on y** is given by

$$b_1 = \frac{SS_{xy}}{SS_y}$$

$$b_0 = \bar{x} - \frac{SS_{xy}}{SS_y}\bar{y} = \bar{x} - b_1\bar{y}.$$

The covariance between X and Y is estimated by $SS_{xy}/(n-1)$, and the correlation by $SS_{xy}/\sqrt{SS_x SS_y}$ .

3. It is instructive to check, in the tables of Appendix 2, that the square of the 97.5th percentile of Student's t with k d.f. is equal to the 95th percentile of F with 1 and k d.f. Similarly, the square of the 95th percentile of t with k d.f. is equal to the 90th percentile of F with 1 and k d.f. Just as $\chi^2$ with 1 d.f. is the square of a normal random variable, so F with 1 and k d.f. is the square of a random variable distributed as t with k d.f. Thus, articles you read in the literature may use either of these equivalent test statistics.

# CHAPTER 10

1. It is possible to calculate the mean squares in Table 10.3 from the data presented in Table 10.2. The mean square among drug groups is 10 (the number of patients in each group) times the variance of the four group means (i.e., 10 times the variance of the set of four numbers 80, 94, 92, and 90). The mean square within drug groups is the pooled within-group variance. Since each group contains the same number of patients, this is the simple average of the four variances, i.e.,

$$\frac{(10.5)^2 + (9.5)^2 + (9.7)^2 + (10.2)^2}{4}$$

2. There are several so-called **multiple comparison** procedures that can be used to test which pairs of a set of means are significantly different. The simplest (Fisher's least significant difference method) is to perform Student's t-test on all the pairwise comparisons, but to require in addition, before making the pairwise comparisons, a significant overall F-test for the equality of the means being considered. The t-tests are performed using the pooled estimate of $\sigma^2$ given by the error mean square of the analysis of variance. Another procedure (the Fisher-Bonferroni method) similarly uses multiple t-tests, but each test must reach significance at the $\alpha/c$ level, where c is the total number of comparisons made, to insure an overall significance level of $\alpha$.

Several procedures (Tukey's, Newman-Keuls', and Duncan's) begin by comparing the largest mean with the smallest, and continue with the next largest difference, and so on, until either a nonsignificant result is encountered or until all pairwise comparisons have been made. At each step the difference is compared to an appropriate null distribution.

A further method (Scheffé's) allows for testing more complex contrasts (such as $H_0$: $\mu_1 - \mu_2 + \mu_3 - \mu_4 = 0$), as well as all pairwise comparisons of means. Finally, there is a procedure (Dunnett's) aimed specifically at identifying group means that are significantly different from the mean of a control group.

All these tests are aimed at maintaining the overall significance level at some fixed value, $\alpha$ (i.e., ensuring that, if the null hypothesis is true, the probability of one or more significant differences being found is $\alpha$).

3. We now give simple rules for writing down formulae for sums of squares that apply in the case of any balanced design (i.e., any design in which all groups of a particular type are the same size). The rules are easy to apply if the degrees of freedom are known for each source of variation. Consider first the "among groups" sums of squares in Table 10.1.

1. The degrees of freedom are $a - 1$.
2. Write down a pair of parentheses and a square (superscript 2) for each term in the degrees of freedom: $(\quad)^2 (\quad)^2$.
3. In front of each pair of parentheses write the appropriate sign corresponding to the term in the degrees-of-freedom, and a summation sign for each letter in the degrees of freedom:

$$\sum_{i=1}^{a} (\quad)^2 - (\quad)^2$$

4. Put summation signs inside the parentheses for all factors and replicates that are not summed over outside the parentheses, followed by the symbol for a typical observation:

$$\sum_{i=1}^{a} \left( \sum_{k=1}^{n} y_{ik} \right)^2 - \left( \sum_{i=1}^{a} \sum_{k=1}^{n} y_{ik} \right)^2$$

5. Finally, divide each term by the number of observations summed within the parentheses. Thus

$$SS_A = \frac{\sum_{i=1}^{a} \left( \sum_{k=1}^{n} y_{ik} \right)^2}{n} - \frac{\left( \sum_{i=1}^{a} \sum_{k=1}^{n} y_{ik} \right)^2}{an}$$

Applying this same sequence of rules, we obtain the sum of squares within groups for Table 10.1, noting that we multiply out the expression for the degrees of freedom.

1. $an - a$

2. $(\quad)^2 (\quad)^2$

3. $\displaystyle\sum_{i=1}^{a} \sum_{k=1}^{n} (\quad)^2 - \sum_{i=1}^{a} (\quad)^2$

4. $\displaystyle\sum_{i=1}^{a} \sum_{k=1}^{n} (y_{ik})^2 - \sum_{i=1}^{a} \left(\sum_{k=1}^{n} y_{ik}\right)^2$

5. $\displaystyle SS_R = \frac{\displaystyle\sum_{i=1}^{a} \sum_{k=1}^{n} (y_{ik})^2}{1} - \frac{\displaystyle\sum_{i=1}^{a} \left(\sum_{k=1}^{n} y_{ik}\right)^2}{n}$

Similarly, for the nested factor analysis of variance in Table 10.4 we obtain:

$$SS_A = \frac{\displaystyle\sum_{i=1}^{a} \left(\sum_{j=1}^{b} \sum_{k=1}^{n} y_{ijk}\right)^2}{bn} - \frac{\left(\displaystyle\sum_{i=1}^{a} \sum_{j=1}^{b} \sum_{b=1}^{n} y_{ijk}\right)^2}{abn}$$

$$SS_B = \frac{\displaystyle\sum_{i=1}^{a} \sum_{j=1}^{b} \left(\sum_{k=1}^{n} y_{ijk}\right)^2}{n} - \frac{\displaystyle\sum_{i=1}^{a} \left(\sum_{j=1}^{b} \sum_{k=1}^{n} y_{ijk}\right)^2}{bn}$$

$$SS_R = \sum_{i=1}^{a} \sum_{j=1}^{b} \sum_{k=1}^{n} y_{ijk}{}^2 - \frac{\displaystyle\sum_{i=1}^{a} \sum_{j=1}^{b} \left(\sum_{k=1}^{n} y_{ijk}\right)^2}{n}$$

For the two-way analysis of variance in Table 10.6 we obtain

$$SS_A = \frac{\displaystyle\sum_{i=1}^{a} \left(\sum_{j=1}^{b} \sum_{k=1}^{n} y_{ijk}\right)^2}{bn} - \frac{\left(\displaystyle\sum_{i=1}^{a} \sum_{j=1}^{b} \sum_{k=1}^{n} y_{ijk}\right)^2}{abn}$$

$$SS_B = \frac{\displaystyle\sum_{j=1}^{b} \left(\sum_{i=1}^{a} \sum_{k=1}^{n} y_{ijk}\right)^2}{an} - \frac{\left(\displaystyle\sum_{i=1}^{a} \sum_{j=1}^{b} \sum_{k=1}^{n} y_{ijk}\right)^2}{abn}$$

$$SS_{AB} = \frac{\displaystyle\sum_{i=1}^{a} \sum_{j=1}^{b} \left(\sum_{k=1}^{n} y_{ijk}\right)^2}{n} - \frac{\displaystyle\sum_{i=1}^{a} \left(\sum_{j=1}^{b} \sum_{k=1}^{n} y_{ijk}\right)^2}{bn}$$

$$- \frac{\sum\limits_{j=1}^{b}\left(\sum\limits_{i=1}^{a}\sum\limits_{k=1}^{n} y_{ijk}\right)^2}{an} + \frac{\left(\sum\limits_{i=1}^{a}\sum\limits_{j=1}^{b}\sum\limits_{k=1}^{n} y_{ijk}\right)^2}{abn}$$

$$SS_R = \sum\limits_{i=1}^{a}\sum\limits_{j=1}^{b}\sum\limits_{k=1}^{n} y_{ijk}^2 - \frac{\sum\limits_{i=1}^{a}\sum\limits_{j=1}^{b}\left(\sum\limits_{k=1}^{n} y_{ijk}\right)^2}{n}$$

# Statistical Tables

Table 1. Percentiles of Student's t-distribution*

| d.f. | $t_{60}$ $-t_{40}$ | $t_{70}$ $-t_{30}$ | $t_{80}$ $-t_{20}$ | $t_{90}$ $-t_{10}$ | $t_{95}$ $-t_5$ | $t_{97.5}$ $-t_{2.5}$ | $t_{99}$ $-t_1$ | $t_{99.5}$ $-t_{0.5}$ |
|---|---|---|---|---|---|---|---|---|
| 1 | .325 | .727 | 1.376 | 3.078 | 6.314 | 12.706 | 31.821 | 63.657 |
| 2 | .289 | .617 | 1.061 | 1.886 | 2.920 | 4.303 | 6.965 | 9.925 |
| 3 | .277 | .584 | .978 | 1.638 | 2.353 | 3.182 | 4.541 | 5.841 |
| 4 | .271 | .569 | .941 | 1.533 | 2.132 | 2.776 | 3.747 | 4.604 |
| 5 | .267 | .559 | .920 | 1.476 | 2.015 | 2.571 | 3.365 | 4.032 |
| 6 | .265 | .553 | .906 | 1.440 | 1.943 | 2.447 | 3.143 | 3.707 |
| 7 | .263 | .549 | .896 | 1.415 | 1.895 | 2.365 | 2.998 | 3.499 |
| 8 | .262 | .546 | .889 | 1.397 | 1.860 | 2.306 | 2.896 | 3.355 |
| 9 | .261 | .543 | .883 | 1.383 | 1.833 | 2.262 | 2.821 | 3.250 |
| 10 | .260 | .542 | .879 | 1.372 | 1.812 | 2.228 | 2.764 | 3.169 |
| 11 | .260 | .540 | .876 | 1.363 | 1.796 | 2.201 | 2.718 | 3.106 |
| 12 | .259 | .539 | .873 | 1.356 | 1.782 | 2.179 | 2.681 | 3.055 |
| 13 | .259 | .538 | .870 | 1.350 | 1.771 | 2.160 | 2.650 | 3.012 |
| 14 | .258 | .537 | .868 | 1.345 | 1.761 | 2.145 | 2.624 | 2.977 |
| 15 | .258 | .536 | .866 | 1.341 | 1.753 | 2.131 | 2.602 | 2.947 |
| 16 | .258 | .535 | .865 | 1.337 | 1.746 | 2.120 | 2.583 | 2.921 |
| 17 | .257 | .534 | .863 | 1.333 | 1.740 | 2.110 | 2.567 | 2.898 |
| 18 | .257 | .534 | .862 | 1.330 | 1.734 | 2.101 | 2.552 | 2.878 |
| 19 | .257 | .533 | .861 | 1.328 | 1.729 | 2.093 | 2.539 | 2.861 |
| 20 | .257 | .533 | .860 | 1.325 | 1.725 | 2.086 | 2.528 | 2.845 |
| 21 | .257 | .532 | .859 | 1.323 | 1.721 | 2.080 | 2.518 | 2.831 |
| 22 | .256 | .532 | .858 | 1.321 | 1.717 | 2.074 | 2.508 | 2.819 |
| 23 | .256 | .532 | .858 | 1.319 | 1.714 | 2.069 | 2.500 | 2.807 |
| 24 | .256 | .531 | .857 | 1.318 | 1.711 | 2.064 | 2.492 | 2.797 |
| 25 | .256 | .531 | .856 | 1.316 | 1.708 | 2.060 | 2.485 | 2.787 |
| 26 | .256 | .531 | .856 | 1.315 | 1.706 | 2.056 | 2.479 | 2.779 |
| 27 | .256 | .531 | .855 | 1.314 | 1.703 | 2.052 | 2.473 | 2.771 |
| 28 | .256 | .530 | .855 | 1.313 | 1.701 | 2.048 | 2.467 | 2.763 |
| 29 | .256 | .530 | .854 | 1.311 | 1.699 | 2.045 | 2.462 | 2.756 |
| 30 | .256 | .530 | .854 | 1.310 | 1.697 | 2.042 | 2.457 | 2.750 |
| 40 | .255 | .529 | .851 | 1.303 | 1.684 | 2.021 | 2.423 | 2.704 |
| 60 | .254 | .527 | .848 | 1.296 | 1.671 | 2.000 | 2.390 | 2.660 |
| 120 | .254 | .526 | .845 | 1.289 | 1.658 | 1.980 | 2.358 | 2.617 |
| ∞ | .253 | .524 | .842 | 1.282 | 1.645 | 1.960 | 2.326 | 2.576 |

*Adapted from Table III of Fisher and Yates: <u>Statistical Tables for Biological, Agricultural and Medical Research</u>, published by Longman Group, Ltd., London (previously published by Oliver and Body, Edinburgh), with permission.

## Table 2. Percentage points of the F-distribution*
### 90th percentiles

| $n_2$ \ $n_1$ | 1 | 2 | 3 | 4 | 5 | 6 | 7 | 8 | 9 | 10 | 12 | 15 | 20 | 24 | 30 | 40 | 60 | 120 | ∞ |
|---|---|---|---|---|---|---|---|---|---|---|---|---|---|---|---|---|---|---|---|
| 1 | 39·86 | 49·50 | 53·59 | 55·83 | 57·24 | 58·20 | 58·91 | 59·44 | 59·86 | 60·19 | 60·71 | 61·22 | 61·74 | 62·00 | 62·26 | 62·53 | 62·79 | 63·06 | 63·33 |
| 2 | 8·53 | 9·00 | 9·16 | 9·24 | 9·29 | 9·33 | 9·35 | 9·37 | 9·38 | 9·39 | 9·41 | 9·42 | 9·44 | 9·45 | 9·46 | 9·47 | 9·47 | 9·48 | 9·49 |
| 3 | 5·54 | 5·46 | 5·39 | 5·34 | 5·31 | 5·28 | 5·27 | 5·25 | 5·24 | 5·23 | 5·22 | 5·20 | 5·18 | 5·18 | 5·17 | 5·16 | 5·15 | 5·14 | 5·13 |
| 4 | 4·54 | 4·32 | 4·19 | 4·11 | 4·05 | 4·01 | 3·98 | 3·95 | 3·94 | 3·92 | 3·90 | 3·87 | 3·84 | 3·83 | 3·82 | 3·80 | 3·79 | 3·78 | 3·76 |
| 5 | 4·06 | 3·78 | 3·62 | 3·52 | 3·45 | 3·40 | 3·37 | 3·34 | 3·32 | 3·30 | 3·27 | 3·24 | 3·21 | 3·19 | 3·17 | 3·16 | 3·14 | 3·12 | 3·10 |
| 6 | 3·78 | 3·46 | 3·29 | 3·18 | 3·11 | 3·05 | 3·01 | 2·98 | 2·96 | 2·94 | 2·90 | 2·87 | 2·84 | 2·82 | 2·80 | 2·78 | 2·76 | 2·74 | 2·72 |
| 7 | 3·59 | 3·26 | 3·07 | 2·96 | 2·88 | 2·83 | 2·78 | 2·75 | 2·72 | 2·70 | 2·67 | 2·63 | 2·59 | 2·58 | 2·56 | 2·54 | 2·51 | 2·49 | 2·47 |
| 8 | 3·46 | 3·11 | 2·92 | 2·81 | 2·73 | 2·67 | 2·62 | 2·59 | 2·56 | 2·54 | 2·50 | 2·46 | 2·42 | 2·40 | 2·38 | 2·36 | 2·34 | 2·32 | 2·29 |
| 9 | 3·36 | 3·01 | 2·81 | 2·69 | 2·61 | 2·55 | 2·51 | 2·47 | 2·44 | 2·42 | 2·38 | 2·34 | 2·30 | 2·28 | 2·25 | 2·23 | 2·21 | 2·18 | 2·16 |
| 10 | 3·29 | 2·92 | 2·73 | 2·61 | 2·52 | 2·46 | 2·41 | 2·38 | 2·35 | 2·32 | 2·28 | 2·24 | 2·20 | 2·18 | 2·16 | 2·13 | 2·11 | 2·08 | 2·06 |
| 11 | 3·23 | 2·86 | 2·66 | 2·54 | 2·45 | 2·39 | 2·34 | 2·30 | 2·27 | 2·25 | 2·21 | 2·17 | 2·12 | 2·10 | 2·08 | 2·05 | 2·03 | 2·00 | 1·97 |
| 12 | 3·18 | 2·81 | 2·61 | 2·48 | 2·39 | 2·33 | 2·28 | 2·24 | 2·21 | 2·19 | 2·15 | 2·10 | 2·06 | 2·04 | 2·01 | 1·99 | 1·96 | 1·93 | 1·90 |
| 13 | 3·14 | 2·76 | 2·56 | 2·43 | 2·35 | 2·28 | 2·23 | 2·20 | 2·16 | 2·14 | 2·10 | 2·05 | 2·01 | 1·98 | 1·96 | 1·93 | 1·90 | 1·88 | 1·85 |
| 14 | 3·10 | 2·73 | 2·52 | 2·39 | 2·31 | 2·24 | 2·19 | 2·15 | 2·12 | 2·10 | 2·05 | 2·01 | 1·96 | 1·94 | 1·91 | 1·89 | 1·86 | 1·83 | 1·80 |
| 15 | 3·07 | 2·70 | 2·49 | 2·36 | 2·27 | 2·21 | 2·16 | 2·12 | 2·09 | 2·06 | 2·02 | 1·97 | 1·92 | 1·90 | 1·87 | 1·85 | 1·82 | 1·79 | 1·76 |
| 16 | 3·05 | 2·67 | 2·46 | 2·33 | 2·24 | 2·18 | 2·13 | 2·09 | 2·06 | 2·03 | 1·99 | 1·94 | 1·89 | 1·87 | 1·84 | 1·81 | 1·78 | 1·75 | 1·72 |
| 17 | 3·03 | 2·64 | 2·44 | 2·31 | 2·22 | 2·15 | 2·10 | 2·06 | 2·03 | 2·00 | 1·96 | 1·91 | 1·86 | 1·84 | 1·81 | 1·78 | 1·75 | 1·72 | 1·69 |
| 18 | 3·01 | 2·62 | 2·42 | 2·29 | 2·20 | 2·13 | 2·08 | 2·04 | 2·00 | 1·98 | 1·93 | 1·89 | 1·84 | 1·81 | 1·78 | 1·75 | 1·72 | 1·69 | 1·66 |
| 19 | 2·99 | 2·61 | 2·40 | 2·27 | 2·18 | 2·11 | 2·06 | 2·02 | 1·98 | 1·96 | 1·91 | 1·86 | 1·81 | 1·79 | 1·76 | 1·73 | 1·70 | 1·67 | 1·63 |
| 20 | 2·97 | 2·59 | 2·38 | 2·25 | 2·16 | 2·09 | 2·04 | 2·00 | 1·96 | 1·94 | 1·89 | 1·84 | 1·79 | 1·77 | 1·74 | 1·71 | 1·68 | 1·64 | 1·61 |
| 21 | 2·96 | 2·57 | 2·36 | 2·23 | 2·14 | 2·08 | 2·02 | 1·98 | 1·95 | 1·92 | 1·87 | 1·83 | 1·78 | 1·75 | 1·72 | 1·69 | 1·66 | 1·62 | 1·59 |
| 22 | 2·95 | 2·56 | 2·35 | 2·22 | 2·13 | 2·06 | 2·01 | 1·97 | 1·93 | 1·90 | 1·86 | 1·81 | 1·76 | 1·73 | 1·70 | 1·67 | 1·64 | 1·60 | 1·57 |
| 23 | 2·94 | 2·55 | 2·34 | 2·21 | 2·11 | 2·05 | 1·99 | 1·95 | 1·92 | 1·89 | 1·84 | 1·80 | 1·74 | 1·72 | 1·69 | 1·66 | 1·62 | 1·59 | 1·55 |
| 24 | 2·93 | 2·54 | 2·33 | 2·19 | 2·10 | 2·04 | 1·98 | 1·94 | 1·91 | 1·88 | 1·83 | 1·78 | 1·73 | 1·70 | 1·67 | 1·64 | 1·61 | 1·57 | 1·53 |
| 25 | 2·92 | 2·53 | 2·32 | 2·18 | 2·09 | 2·02 | 1·97 | 1·93 | 1·89 | 1·87 | 1·82 | 1·77 | 1·72 | 1·69 | 1·66 | 1·63 | 1·59 | 1·56 | 1·52 |
| 26 | 2·91 | 2·52 | 2·31 | 2·17 | 2·08 | 2·01 | 1·96 | 1·92 | 1·88 | 1·86 | 1·81 | 1·76 | 1·71 | 1·68 | 1·65 | 1·61 | 1·58 | 1·54 | 1·50 |
| 27 | 2·90 | 2·51 | 2·30 | 2·17 | 2·07 | 2·00 | 1·95 | 1·91 | 1·87 | 1·85 | 1·80 | 1·75 | 1·70 | 1·67 | 1·64 | 1·60 | 1·57 | 1·53 | 1·49 |
| 28 | 2·89 | 2·50 | 2·29 | 2·16 | 2·06 | 2·00 | 1·94 | 1·90 | 1·87 | 1·84 | 1·79 | 1·74 | 1·69 | 1·66 | 1·63 | 1·59 | 1·56 | 1·52 | 1·48 |
| 29 | 2·89 | 2·50 | 2·28 | 2·15 | 2·06 | 1·99 | 1·93 | 1·89 | 1·86 | 1·83 | 1·78 | 1·73 | 1·68 | 1·65 | 1·62 | 1·58 | 1·55 | 1·51 | 1·47 |
| 30 | 2·88 | 2·49 | 2·28 | 2·14 | 2·05 | 1·98 | 1·93 | 1·88 | 1·85 | 1·82 | 1·77 | 1·72 | 1·67 | 1·64 | 1·61 | 1·57 | 1·54 | 1·50 | 1·46 |
| 40 | 2·84 | 2·44 | 2·23 | 2·09 | 2·00 | 1·93 | 1·87 | 1·83 | 1·79 | 1·76 | 1·71 | 1·66 | 1·61 | 1·57 | 1·54 | 1·51 | 1·47 | 1·42 | 1·38 |
| 60 | 2·79 | 2·39 | 2·18 | 2·04 | 1·95 | 1·87 | 1·82 | 1·77 | 1·74 | 1·71 | 1·66 | 1·60 | 1·54 | 1·51 | 1·48 | 1·44 | 1·40 | 1·35 | 1·29 |
| 120 | 2·75 | 2·35 | 2·13 | 1·99 | 1·90 | 1·82 | 1·77 | 1·72 | 1·68 | 1·65 | 1·60 | 1·55 | 1·48 | 1·45 | 1·41 | 1·37 | 1·32 | 1·26 | 1·19 |
| ∞ | 2·71 | 2·30 | 2·08 | 1·94 | 1·85 | 1·77 | 1·72 | 1·67 | 1·63 | 1·60 | 1·55 | 1·49 | 1·42 | 1·38 | 1·34 | 1·30 | 1·24 | 1·17 | 1·00 |

*Reprinted from Biometrika Tables for Statisticians (2nd ed., 1962), eds. E. S. Pearson and H. O. Hartley, by permission of the Biometrika Trustees and the publisher, Cambridge University Press.

Table 2. Percentage points of the F-distribution (continued)

95th percentiles

| $n_2$ \ $n_1$ | 1 | 2 | 3 | 4 | 5 | 6 | 7 | 8 | 9 | 10 | 12 | 15 | 20 | 24 | 30 | 40 | 60 | 120 | ∞ |
|---|---|---|---|---|---|---|---|---|---|---|---|---|---|---|---|---|---|---|---|
| 1 | 161.4 | 199.5 | 215.7 | 224.6 | 230.2 | 234.0 | 236.8 | 238.9 | 240.5 | 241.9 | 243.9 | 245.9 | 248.0 | 249.1 | 250.1 | 251.1 | 252.2 | 253.3 | 254.3 |
| 2 | 18.51 | 19.00 | 19.16 | 19.25 | 19.30 | 19.33 | 19.35 | 19.37 | 19.38 | 19.40 | 19.41 | 19.43 | 19.45 | 19.45 | 19.46 | 19.47 | 19.48 | 19.49 | 19.50 |
| 3 | 10.13 | 9.55 | 9.28 | 9.12 | 9.01 | 8.94 | 8.89 | 8.85 | 8.81 | 8.79 | 8.74 | 8.70 | 8.66 | 8.64 | 8.62 | 8.59 | 8.57 | 8.55 | 8.53 |
| 4 | 7.71 | 6.94 | 6.59 | 6.39 | 6.26 | 6.16 | 6.09 | 6.04 | 6.00 | 5.96 | 5.91 | 5.86 | 5.80 | 5.77 | 5.75 | 5.72 | 5.69 | 5.66 | 5.63 |
| 5 | 6.61 | 5.79 | 5.41 | 5.19 | 5.05 | 4.95 | 4.88 | 4.82 | 4.77 | 4.74 | 4.68 | 4.62 | 4.56 | 4.53 | 4.50 | 4.46 | 4.43 | 4.40 | 4.36 |
| 6 | 5.99 | 5.14 | 4.76 | 4.53 | 4.39 | 4.28 | 4.21 | 4.15 | 4.10 | 4.06 | 4.00 | 3.94 | 3.87 | 3.84 | 3.81 | 3.77 | 3.74 | 3.70 | 3.67 |
| 7 | 5.59 | 4.74 | 4.35 | 4.12 | 3.97 | 3.87 | 3.79 | 3.73 | 3.68 | 3.64 | 3.57 | 3.51 | 3.44 | 3.41 | 3.38 | 3.34 | 3.30 | 3.27 | 3.23 |
| 8 | 5.32 | 4.46 | 4.07 | 3.84 | 3.69 | 3.58 | 3.50 | 3.44 | 3.39 | 3.35 | 3.28 | 3.22 | 3.15 | 3.12 | 3.08 | 3.04 | 3.01 | 2.97 | 2.93 |
| 9 | 5.12 | 4.26 | 3.86 | 3.63 | 3.48 | 3.37 | 3.29 | 3.23 | 3.18 | 3.14 | 3.07 | 3.01 | 2.94 | 2.90 | 2.86 | 2.83 | 2.79 | 2.75 | 2.71 |
| 10 | 4.96 | 4.10 | 3.71 | 3.48 | 3.33 | 3.22 | 3.14 | 3.07 | 3.02 | 2.98 | 2.91 | 2.85 | 2.77 | 2.74 | 2.70 | 2.66 | 2.62 | 2.58 | 2.54 |
| 11 | 4.84 | 3.98 | 3.59 | 3.36 | 3.20 | 3.09 | 3.01 | 2.95 | 2.90 | 2.85 | 2.79 | 2.72 | 2.65 | 2.61 | 2.57 | 2.53 | 2.49 | 2.45 | 2.40 |
| 12 | 4.75 | 3.89 | 3.49 | 3.26 | 3.11 | 3.00 | 2.91 | 2.85 | 2.80 | 2.75 | 2.69 | 2.62 | 2.54 | 2.51 | 2.47 | 2.43 | 2.38 | 2.34 | 2.30 |
| 13 | 4.67 | 3.81 | 3.41 | 3.18 | 3.03 | 2.92 | 2.83 | 2.77 | 2.71 | 2.67 | 2.60 | 2.53 | 2.46 | 2.42 | 2.38 | 2.34 | 2.30 | 2.25 | 2.21 |
| 14 | 4.60 | 3.74 | 3.34 | 3.11 | 2.96 | 2.85 | 2.76 | 2.70 | 2.65 | 2.60 | 2.53 | 2.46 | 2.39 | 2.35 | 2.31 | 2.27 | 2.22 | 2.18 | 2.13 |
| 15 | 4.54 | 3.68 | 3.29 | 3.06 | 2.90 | 2.79 | 2.71 | 2.64 | 2.59 | 2.54 | 2.48 | 2.40 | 2.33 | 2.29 | 2.25 | 2.20 | 2.16 | 2.11 | 2.07 |
| 16 | 4.49 | 3.63 | 3.24 | 3.01 | 2.85 | 2.74 | 2.66 | 2.59 | 2.54 | 2.49 | 2.42 | 2.35 | 2.28 | 2.24 | 2.19 | 2.15 | 2.11 | 2.06 | 2.01 |
| 17 | 4.45 | 3.59 | 3.20 | 2.96 | 2.81 | 2.70 | 2.61 | 2.55 | 2.49 | 2.45 | 2.38 | 2.31 | 2.23 | 2.19 | 2.15 | 2.10 | 2.06 | 2.01 | 1.96 |
| 18 | 4.41 | 3.55 | 3.16 | 2.93 | 2.77 | 2.66 | 2.58 | 2.51 | 2.46 | 2.41 | 2.34 | 2.27 | 2.19 | 2.15 | 2.11 | 2.06 | 2.02 | 1.97 | 1.92 |
| 19 | 4.38 | 3.52 | 3.13 | 2.90 | 2.74 | 2.63 | 2.54 | 2.48 | 2.42 | 2.38 | 2.31 | 2.23 | 2.16 | 2.11 | 2.07 | 2.03 | 1.98 | 1.93 | 1.88 |
| 20 | 4.35 | 3.49 | 3.10 | 2.87 | 2.71 | 2.60 | 2.51 | 2.45 | 2.39 | 2.35 | 2.28 | 2.20 | 2.12 | 2.08 | 2.04 | 1.99 | 1.95 | 1.90 | 1.84 |
| 21 | 4.32 | 3.47 | 3.07 | 2.84 | 2.68 | 2.57 | 2.49 | 2.42 | 2.37 | 2.32 | 2.25 | 2.18 | 2.10 | 2.05 | 2.01 | 1.96 | 1.92 | 1.87 | 1.81 |
| 22 | 4.30 | 3.44 | 3.05 | 2.82 | 2.66 | 2.55 | 2.46 | 2.40 | 2.34 | 2.30 | 2.23 | 2.15 | 2.07 | 2.03 | 1.98 | 1.94 | 1.89 | 1.84 | 1.78 |
| 23 | 4.28 | 3.42 | 3.03 | 2.80 | 2.64 | 2.53 | 2.44 | 2.37 | 2.32 | 2.27 | 2.20 | 2.13 | 2.05 | 2.01 | 1.96 | 1.91 | 1.86 | 1.81 | 1.76 |
| 24 | 4.26 | 3.40 | 3.01 | 2.78 | 2.62 | 2.51 | 2.42 | 2.36 | 2.30 | 2.25 | 2.18 | 2.11 | 2.03 | 1.98 | 1.94 | 1.89 | 1.84 | 1.79 | 1.73 |
| 25 | 4.24 | 3.39 | 2.99 | 2.76 | 2.60 | 2.49 | 2.40 | 2.34 | 2.28 | 2.24 | 2.16 | 2.09 | 2.01 | 1.96 | 1.92 | 1.87 | 1.82 | 1.77 | 1.71 |
| 26 | 4.23 | 3.37 | 2.98 | 2.74 | 2.59 | 2.47 | 2.39 | 2.32 | 2.27 | 2.22 | 2.15 | 2.07 | 1.99 | 1.95 | 1.90 | 1.85 | 1.80 | 1.75 | 1.69 |
| 27 | 4.21 | 3.35 | 2.96 | 2.73 | 2.57 | 2.46 | 2.37 | 2.31 | 2.25 | 2.20 | 2.13 | 2.06 | 1.97 | 1.93 | 1.88 | 1.84 | 1.79 | 1.73 | 1.67 |
| 28 | 4.20 | 3.34 | 2.95 | 2.71 | 2.56 | 2.45 | 2.36 | 2.29 | 2.24 | 2.19 | 2.12 | 2.04 | 1.96 | 1.91 | 1.87 | 1.82 | 1.77 | 1.71 | 1.65 |
| 29 | 4.18 | 3.33 | 2.93 | 2.70 | 2.55 | 2.43 | 2.35 | 2.28 | 2.22 | 2.18 | 2.10 | 2.03 | 1.94 | 1.90 | 1.85 | 1.81 | 1.75 | 1.70 | 1.64 |
| 30 | 4.17 | 3.32 | 2.92 | 2.69 | 2.53 | 2.42 | 2.33 | 2.27 | 2.21 | 2.16 | 2.09 | 2.01 | 1.93 | 1.89 | 1.84 | 1.79 | 1.74 | 1.68 | 1.62 |
| 40 | 4.08 | 3.23 | 2.84 | 2.61 | 2.45 | 2.34 | 2.25 | 2.18 | 2.12 | 2.08 | 2.00 | 1.92 | 1.84 | 1.79 | 1.74 | 1.69 | 1.64 | 1.58 | 1.51 |
| 60 | 4.00 | 3.15 | 2.76 | 2.53 | 2.37 | 2.25 | 2.17 | 2.10 | 2.04 | 1.99 | 1.92 | 1.84 | 1.75 | 1.70 | 1.65 | 1.59 | 1.53 | 1.47 | 1.39 |
| 120 | 3.92 | 3.07 | 2.68 | 2.45 | 2.29 | 2.17 | 2.09 | 2.02 | 1.96 | 1.91 | 1.83 | 1.75 | 1.66 | 1.61 | 1.55 | 1.50 | 1.43 | 1.35 | 1.25 |
| ∞ | 3.84 | 3.00 | 2.60 | 2.37 | 2.21 | 2.10 | 2.01 | 1.94 | 1.88 | 1.83 | 1.75 | 1.67 | 1.57 | 1.52 | 1.46 | 1.39 | 1.32 | 1.22 | 1.00 |

Table 2 (continued)
97.5th percentiles

| $n_2$ \ $n_1$ | 1 | 2 | 3 | 4 | 5 | 6 | 7 | 8 | 9 | 10 | 12 | 15 | 20 | 24 | 30 | 40 | 60 | 120 | ∞ |
|---|---|---|---|---|---|---|---|---|---|---|---|---|---|---|---|---|---|---|---|
| 1 | 647.8 | 799.5 | 864.2 | 899.6 | 921.8 | 937.1 | 948.2 | 956.7 | 963.3 | 968.6 | 976.7 | 984.9 | 993.1 | 997.2 | 1001 | 1006 | 1010 | 1014 | 1018 |
| 2 | 38.51 | 39.00 | 39.17 | 39.25 | 39.30 | 39.33 | 39.36 | 39.37 | 39.39 | 39.40 | 39.41 | 39.43 | 39.45 | 39.46 | 39.46 | 39.47 | 39.48 | 39.49 | 39.50 |
| 3 | 17.44 | 16.04 | 15.44 | 15.10 | 14.88 | 14.73 | 14.62 | 14.54 | 14.47 | 14.42 | 14.34 | 14.25 | 14.17 | 14.12 | 14.08 | 14.04 | 13.99 | 13.95 | 13.90 |
| 4 | 12.22 | 10.65 | 9.98 | 9.60 | 9.36 | 9.20 | 9.07 | 8.98 | 8.90 | 8.84 | 8.75 | 8.66 | 8.56 | 8.51 | 8.46 | 8.41 | 8.36 | 8.31 | 8.26 |
| 5 | 10.01 | 8.43 | 7.76 | 7.39 | 7.15 | 6.98 | 6.85 | 6.76 | 6.68 | 6.62 | 6.52 | 6.43 | 6.33 | 6.28 | 6.23 | 6.18 | 6.12 | 6.07 | 6.02 |
| 6 | 8.81 | 7.26 | 6.60 | 6.23 | 5.99 | 5.82 | 5.70 | 5.60 | 5.52 | 5.46 | 5.37 | 5.27 | 5.17 | 5.12 | 5.07 | 5.01 | 4.96 | 4.90 | 4.85 |
| 7 | 8.07 | 6.54 | 5.89 | 5.52 | 5.29 | 5.12 | 4.99 | 4.90 | 4.82 | 4.76 | 4.67 | 4.57 | 4.47 | 4.42 | 4.36 | 4.31 | 4.25 | 4.20 | 4.14 |
| 8 | 7.57 | 6.06 | 5.42 | 5.05 | 4.82 | 4.65 | 4.53 | 4.43 | 4.36 | 4.30 | 4.20 | 4.10 | 4.00 | 3.95 | 3.89 | 3.84 | 3.78 | 3.73 | 3.67 |
| 9 | 7.21 | 5.71 | 5.08 | 4.72 | 4.48 | 4.32 | 4.20 | 4.10 | 4.03 | 3.96 | 3.87 | 3.77 | 3.67 | 3.61 | 3.56 | 3.51 | 3.45 | 3.39 | 3.33 |
| 10 | 6.94 | 5.46 | 4.83 | 4.47 | 4.24 | 4.07 | 3.95 | 3.85 | 3.78 | 3.72 | 3.62 | 3.52 | 3.42 | 3.37 | 3.31 | 3.26 | 3.20 | 3.14 | 3.08 |
| 11 | 6.72 | 5.26 | 4.63 | 4.28 | 4.04 | 3.88 | 3.76 | 3.66 | 3.59 | 3.53 | 3.43 | 3.33 | 3.23 | 3.17 | 3.12 | 3.06 | 3.00 | 2.94 | 2.88 |
| 12 | 6.55 | 5.10 | 4.47 | 4.12 | 3.89 | 3.73 | 3.61 | 3.51 | 3.44 | 3.37 | 3.28 | 3.18 | 3.07 | 3.02 | 2.96 | 2.91 | 2.85 | 2.79 | 2.72 |
| 13 | 6.41 | 4.97 | 4.35 | 4.00 | 3.77 | 3.60 | 3.48 | 3.39 | 3.31 | 3.25 | 3.15 | 3.05 | 2.95 | 2.89 | 2.84 | 2.78 | 2.72 | 2.66 | 2.60 |
| 14 | 6.30 | 4.86 | 4.24 | 3.89 | 3.66 | 3.50 | 3.38 | 3.29 | 3.21 | 3.15 | 3.05 | 2.95 | 2.84 | 2.79 | 2.73 | 2.67 | 2.61 | 2.55 | 2.49 |
| 15 | 6.20 | 4.77 | 4.15 | 3.80 | 3.58 | 3.41 | 3.29 | 3.20 | 3.12 | 3.06 | 2.96 | 2.86 | 2.76 | 2.70 | 2.64 | 2.59 | 2.52 | 2.46 | 2.40 |
| 16 | 6.12 | 4.69 | 4.08 | 3.73 | 3.50 | 3.34 | 3.22 | 3.12 | 3.05 | 2.99 | 2.89 | 2.79 | 2.68 | 2.63 | 2.57 | 2.51 | 2.45 | 2.38 | 2.32 |
| 17 | 6.04 | 4.62 | 4.01 | 3.66 | 3.44 | 3.28 | 3.16 | 3.06 | 2.98 | 2.92 | 2.82 | 2.72 | 2.62 | 2.56 | 2.50 | 2.44 | 2.38 | 2.32 | 2.25 |
| 18 | 5.98 | 4.56 | 3.95 | 3.61 | 3.38 | 3.22 | 3.10 | 3.01 | 2.93 | 2.87 | 2.77 | 2.67 | 2.56 | 2.50 | 2.44 | 2.38 | 2.32 | 2.26 | 2.19 |
| 19 | 5.92 | 4.51 | 3.90 | 3.56 | 3.33 | 3.17 | 3.05 | 2.96 | 2.88 | 2.82 | 2.72 | 2.62 | 2.51 | 2.45 | 2.39 | 2.33 | 2.27 | 2.20 | 2.13 |
| 20 | 5.87 | 4.46 | 3.86 | 3.51 | 3.29 | 3.13 | 3.01 | 2.91 | 2.84 | 2.77 | 2.68 | 2.57 | 2.46 | 2.41 | 2.35 | 2.29 | 2.22 | 2.16 | 2.09 |
| 21 | 5.83 | 4.42 | 3.82 | 3.48 | 3.25 | 3.09 | 2.97 | 2.87 | 2.80 | 2.73 | 2.64 | 2.53 | 2.42 | 2.37 | 2.31 | 2.25 | 2.18 | 2.11 | 2.04 |
| 22 | 5.79 | 4.38 | 3.78 | 3.44 | 3.22 | 3.05 | 2.93 | 2.84 | 2.76 | 2.70 | 2.60 | 2.50 | 2.39 | 2.33 | 2.27 | 2.21 | 2.14 | 2.08 | 2.00 |
| 23 | 5.75 | 4.35 | 3.75 | 3.41 | 3.18 | 3.02 | 2.90 | 2.81 | 2.73 | 2.67 | 2.57 | 2.47 | 2.36 | 2.30 | 2.24 | 2.18 | 2.11 | 2.04 | 1.97 |
| 24 | 5.72 | 4.32 | 3.72 | 3.38 | 3.15 | 2.99 | 2.87 | 2.78 | 2.70 | 2.64 | 2.54 | 2.44 | 2.33 | 2.27 | 2.21 | 2.15 | 2.08 | 2.01 | 1.94 |
| 25 | 5.69 | 4.29 | 3.69 | 3.35 | 3.13 | 2.97 | 2.85 | 2.75 | 2.68 | 2.61 | 2.51 | 2.41 | 2.30 | 2.24 | 2.18 | 2.12 | 2.05 | 1.98 | 1.91 |
| 26 | 5.66 | 4.27 | 3.67 | 3.33 | 3.10 | 2.94 | 2.82 | 2.73 | 2.65 | 2.59 | 2.49 | 2.39 | 2.28 | 2.22 | 2.16 | 2.09 | 2.03 | 1.95 | 1.88 |
| 27 | 5.63 | 4.24 | 3.65 | 3.31 | 3.08 | 2.92 | 2.80 | 2.71 | 2.63 | 2.57 | 2.47 | 2.36 | 2.25 | 2.19 | 2.13 | 2.07 | 2.00 | 1.93 | 1.85 |
| 28 | 5.61 | 4.22 | 3.63 | 3.29 | 3.06 | 2.90 | 2.78 | 2.69 | 2.61 | 2.55 | 2.45 | 2.34 | 2.23 | 2.17 | 2.11 | 2.05 | 1.98 | 1.91 | 1.83 |
| 29 | 5.59 | 4.20 | 3.61 | 3.27 | 3.04 | 2.88 | 2.76 | 2.67 | 2.59 | 2.53 | 2.43 | 2.32 | 2.21 | 2.15 | 2.09 | 2.03 | 1.96 | 1.89 | 1.81 |
| 30 | 5.57 | 4.18 | 3.59 | 3.25 | 3.03 | 2.87 | 2.75 | 2.65 | 2.57 | 2.51 | 2.41 | 2.31 | 2.20 | 2.14 | 2.07 | 2.01 | 1.94 | 1.87 | 1.79 |
| 40 | 5.42 | 4.05 | 3.46 | 3.13 | 2.90 | 2.74 | 2.62 | 2.53 | 2.45 | 2.39 | 2.29 | 2.18 | 2.07 | 2.01 | 1.94 | 1.88 | 1.80 | 1.72 | 1.64 |
| 60 | 5.29 | 3.93 | 3.34 | 3.01 | 2.79 | 2.63 | 2.51 | 2.41 | 2.33 | 2.27 | 2.17 | 2.06 | 1.94 | 1.88 | 1.82 | 1.74 | 1.67 | 1.58 | 1.48 |
| 120 | 5.15 | 3.80 | 3.23 | 2.89 | 2.67 | 2.52 | 2.39 | 2.30 | 2.22 | 2.16 | 2.05 | 1.94 | 1.82 | 1.76 | 1.69 | 1.61 | 1.53 | 1.43 | 1.31 |
| ∞ | 5.02 | 3.69 | 3.12 | 2.79 | 2.57 | 2.41 | 2.29 | 2.19 | 2.11 | 2.05 | 1.94 | 1.83 | 1.71 | 1.64 | 1.57 | 1.48 | 1.39 | 1.27 | 1.00 |

Table 2. Percentage points of the F-distribution (continued)

99th percentiles

| $n_2$ \ $n_1$ | 1 | 2 | 3 | 4 | 5 | 6 | 7 | 8 | 9 | 10 | 12 | 15 | 20 | 24 | 30 | 40 | 60 | 120 | ∞ |
|---|---|---|---|---|---|---|---|---|---|---|---|---|---|---|---|---|---|---|---|
| 1 | 4052 | 4999·5 | 5403 | 5625 | 5764 | 5859 | 5928 | 5982 | 6022 | 6056 | 6106 | 6157 | 6209 | 6235 | 6261 | 6287 | 6313 | 6339 | 6366 |
| 2 | 98·50 | 99·00 | 99·17 | 99·25 | 99·30 | 99·33 | 99·36 | 99·37 | 99·39 | 99·40 | 99·42 | 99·43 | 99·45 | 99·46 | 99·47 | 99·47 | 99·48 | 99·49 | 99·50 |
| 3 | 34·12 | 30·82 | 29·46 | 28·71 | 28·24 | 27·91 | 27·67 | 27·49 | 27·35 | 27·23 | 27·05 | 26·87 | 26·69 | 26·60 | 26·50 | 26·41 | 26·32 | 26·22 | 26·13 |
| 4 | 21·20 | 18·00 | 16·69 | 15·98 | 15·52 | 15·21 | 14·98 | 14·80 | 14·66 | 14·55 | 14·37 | 14·20 | 14·02 | 13·93 | 13·84 | 13·75 | 13·65 | 13·56 | 13·46 |
| 5 | 16·26 | 13·27 | 12·06 | 11·39 | 10·97 | 10·67 | 10·46 | 10·29 | 10·16 | 10·05 | 9·89 | 9·72 | 9·55 | 9·47 | 9·38 | 9·29 | 9·20 | 9·11 | 9·02 |
| 6 | 13·75 | 10·92 | 9·78 | 9·15 | 8·75 | 8·47 | 8·26 | 8·10 | 7·98 | 7·87 | 7·72 | 7·56 | 7·40 | 7·31 | 7·23 | 7·14 | 7·06 | 6·97 | 6·88 |
| 7 | 12·25 | 9·55 | 8·45 | 7·85 | 7·46 | 7·19 | 6·99 | 6·84 | 6·72 | 6·62 | 6·47 | 6·31 | 6·16 | 6·07 | 5·99 | 5·91 | 5·82 | 5·74 | 5·65 |
| 8 | 11·26 | 8·65 | 7·59 | 7·01 | 6·63 | 6·37 | 6·18 | 6·03 | 5·91 | 5·81 | 5·67 | 5·52 | 5·36 | 5·28 | 5·20 | 5·12 | 5·03 | 4·95 | 4·86 |
| 9 | 10·56 | 8·02 | 6·99 | 6·42 | 6·06 | 5·80 | 5·61 | 5·47 | 5·35 | 5·26 | 5·11 | 4·96 | 4·81 | 4·73 | 4·65 | 4·57 | 4·48 | 4·40 | 4·31 |
| 10 | 10·04 | 7·56 | 6·55 | 5·99 | 5·64 | 5·39 | 5·20 | 5·06 | 4·94 | 4·85 | 4·71 | 4·56 | 4·41 | 4·33 | 4·25 | 4·17 | 4·08 | 4·00 | 3·91 |
| 11 | 9·65 | 7·21 | 6·22 | 5·67 | 5·32 | 5·07 | 4·89 | 4·74 | 4·63 | 4·54 | 4·40 | 4·25 | 4·10 | 4·02 | 3·94 | 3·86 | 3·78 | 3·69 | 3·60 |
| 12 | 9·33 | 6·93 | 5·95 | 5·41 | 5·06 | 4·82 | 4·64 | 4·50 | 4·39 | 4·30 | 4·16 | 4·01 | 3·86 | 3·78 | 3·70 | 3·62 | 3·54 | 3·45 | 3·36 |
| 13 | 9·07 | 6·70 | 5·74 | 5·21 | 4·86 | 4·62 | 4·44 | 4·30 | 4·19 | 4·10 | 3·96 | 3·82 | 3·66 | 3·59 | 3·51 | 3·43 | 3·34 | 3·25 | 3·17 |
| 14 | 8·86 | 6·51 | 5·56 | 5·04 | 4·69 | 4·46 | 4·28 | 4·14 | 4·03 | 3·94 | 3·80 | 3·66 | 3·51 | 3·43 | 3·35 | 3·27 | 3·18 | 3·09 | 3·00 |
| 15 | 8·68 | 6·36 | 5·42 | 4·89 | 4·56 | 4·32 | 4·14 | 4·00 | 3·89 | 3·80 | 3·67 | 3·52 | 3·37 | 3·29 | 3·21 | 3·13 | 3·05 | 2·96 | 2·87 |
| 16 | 8·53 | 6·23 | 5·29 | 4·77 | 4·44 | 4·20 | 4·03 | 3·89 | 3·78 | 3·69 | 3·55 | 3·41 | 3·26 | 3·18 | 3·10 | 3·02 | 2·93 | 2·84 | 2·75 |
| 17 | 8·40 | 6·11 | 5·18 | 4·67 | 4·34 | 4·10 | 3·93 | 3·79 | 3·68 | 3·59 | 3·46 | 3·31 | 3·16 | 3·08 | 3·00 | 2·92 | 2·83 | 2·75 | 2·65 |
| 18 | 8·29 | 6·01 | 5·09 | 4·58 | 4·25 | 4·01 | 3·84 | 3·71 | 3·60 | 3·51 | 3·37 | 3·23 | 3·08 | 3·00 | 2·92 | 2·84 | 2·75 | 2·66 | 2·57 |
| 19 | 8·18 | 5·93 | 5·01 | 4·50 | 4·17 | 3·94 | 3·77 | 3·63 | 3·52 | 3·43 | 3·30 | 3·15 | 3·00 | 2·92 | 2·84 | 2·76 | 2·67 | 2·58 | 2·49 |
| 20 | 8·10 | 5·85 | 4·94 | 4·43 | 4·10 | 3·87 | 3·70 | 3·56 | 3·46 | 3·37 | 3·23 | 3·09 | 2·94 | 2·86 | 2·78 | 2·69 | 2·61 | 2·52 | 2·42 |
| 21 | 8·02 | 5·78 | 4·87 | 4·37 | 4·04 | 3·81 | 3·64 | 3·51 | 3·40 | 3·31 | 3·17 | 3·03 | 2·88 | 2·80 | 2·72 | 2·64 | 2·55 | 2·46 | 2·36 |
| 22 | 7·95 | 5·72 | 4·82 | 4·31 | 3·99 | 3·76 | 3·59 | 3·45 | 3·35 | 3·26 | 3·12 | 2·98 | 2·83 | 2·75 | 2·67 | 2·58 | 2·50 | 2·40 | 2·31 |
| 23 | 7·88 | 5·66 | 4·76 | 4·26 | 3·94 | 3·71 | 3·54 | 3·41 | 3·30 | 3·21 | 3·07 | 2·93 | 2·78 | 2·70 | 2·62 | 2·54 | 2·45 | 2·35 | 2·26 |
| 24 | 7·82 | 5·61 | 4·72 | 4·22 | 3·90 | 3·67 | 3·50 | 3·36 | 3·26 | 3·17 | 3·03 | 2·89 | 2·74 | 2·66 | 2·58 | 2·49 | 2·40 | 2·31 | 2·21 |
| 25 | 7·77 | 5·57 | 4·68 | 4·18 | 3·85 | 3·63 | 3·46 | 3·32 | 3·22 | 3·13 | 2·99 | 2·85 | 2·70 | 2·62 | 2·54 | 2·45 | 2·36 | 2·27 | 2·17 |
| 26 | 7·72 | 5·53 | 4·64 | 4·14 | 3·82 | 3·59 | 3·42 | 3·29 | 3·18 | 3·09 | 2·96 | 2·81 | 2·66 | 2·58 | 2·50 | 2·42 | 2·33 | 2·23 | 2·13 |
| 27 | 7·68 | 5·49 | 4·60 | 4·11 | 3·78 | 3·56 | 3·39 | 3·26 | 3·15 | 3·06 | 2·93 | 2·78 | 2·63 | 2·55 | 2·47 | 2·38 | 2·29 | 2·20 | 2·10 |
| 28 | 7·64 | 5·45 | 4·57 | 4·07 | 3·75 | 3·53 | 3·36 | 3·23 | 3·12 | 3·03 | 2·90 | 2·75 | 2·60 | 2·52 | 2·44 | 2·35 | 2·26 | 2·17 | 2·06 |
| 29 | 7·60 | 5·42 | 4·54 | 4·04 | 3·73 | 3·50 | 3·33 | 3·20 | 3·09 | 3·00 | 2·87 | 2·73 | 2·57 | 2·49 | 2·41 | 2·33 | 2·23 | 2·14 | 2·03 |
| 30 | 7·56 | 5·39 | 4·51 | 4·02 | 3·70 | 3·47 | 3·30 | 3·17 | 3·07 | 2·98 | 2·84 | 2·70 | 2·55 | 2·47 | 2·39 | 2·30 | 2·21 | 2·11 | 2·01 |
| 40 | 7·31 | 5·18 | 4·31 | 3·83 | 3·51 | 3·29 | 3·12 | 2·99 | 2·89 | 2·80 | 2·66 | 2·52 | 2·37 | 2·29 | 2·20 | 2·11 | 2·02 | 1·92 | 1·80 |
| 60 | 7·08 | 4·98 | 4·13 | 3·65 | 3·34 | 3·12 | 2·95 | 2·82 | 2·72 | 2·63 | 2·50 | 2·35 | 2·20 | 2·12 | 2·03 | 1·94 | 1·84 | 1·73 | 1·60 |
| 120 | 6·85 | 4·79 | 3·95 | 3·48 | 3·17 | 2·96 | 2·79 | 2·66 | 2·56 | 2·47 | 2·34 | 2·19 | 2·03 | 1·95 | 1·86 | 1·76 | 1·66 | 1·53 | 1·38 |
| ∞ | 6·63 | 4·61 | 3·78 | 3·32 | 3·02 | 2·80 | 2·64 | 2·51 | 2·41 | 2·32 | 2·18 | 2·04 | 1·88 | 1·79 | 1·70 | 1·59 | 1·47 | 1·32 | 1·00 |

Table 2 (continued)
99.5th percentiles

| $n_2$ \ $n_1$ | 1 | 2 | 3 | 4 | 5 | 6 | 7 | 8 | 9 | 10 | 12 | 15 | 20 | 24 | 30 | 40 | 60 | 120 | $\infty$ |
|---|---|---|---|---|---|---|---|---|---|---|---|---|---|---|---|---|---|---|---|
| 1 | 16211 | 20000 | 21615 | 22500 | 23056 | 23437 | 23715 | 23925 | 24091 | 24224 | 24426 | 24630 | 24836 | 24940 | 25044 | 25148 | 25253 | 25359 | 25465 |
| 2 | 198·5 | 199·0 | 199·2 | 199·2 | 199·3 | 199·3 | 199·4 | 199·4 | 199·4 | 199·4 | 199·4 | 199·4 | 199·4 | 199·5 | 199·5 | 199·5 | 199·5 | 199·5 | 199·5 |
| 3 | 55·55 | 49·80 | 47·47 | 46·19 | 45·39 | 44·84 | 44·43 | 44·13 | 43·88 | 43·69 | 43·39 | 43·08 | 42·78 | 42·62 | 42·47 | 42·31 | 42·15 | 41·99 | 41·83 |
| 4 | 31·33 | 26·28 | 24·26 | 23·15 | 22·46 | 21·97 | 21·62 | 21·35 | 21·14 | 20·97 | 20·70 | 20·44 | 20·17 | 20·03 | 19·89 | 19·75 | 19·61 | 19·47 | 19·32 |
| 5 | 22·78 | 18·31 | 16·53 | 15·56 | 14·94 | 14·51 | 14·20 | 13·96 | 13·77 | 13·62 | 13·38 | 13·15 | 12·90 | 12·78 | 12·66 | 12·53 | 12·40 | 12·27 | 12·14 |
| 6 | 18·63 | 14·54 | 12·92 | 12·03 | 11·46 | 11·07 | 10·79 | 10·57 | 10·39 | 10·25 | 10·03 | 9·81 | 9·59 | 9·47 | 9·36 | 9·24 | 9·12 | 9·00 | 8·88 |
| 7 | 16·24 | 12·40 | 10·88 | 10·05 | 9·52 | 9·16 | 8·89 | 8·68 | 8·51 | 8·38 | 8·18 | 7·97 | 7·75 | 7·65 | 7·53 | 7·42 | 7·31 | 7·19 | 7·08 |
| 8 | 14·69 | 11·04 | 9·60 | 8·81 | 8·30 | 7·95 | 7·69 | 7·50 | 7·34 | 7·21 | 7·01 | 6·81 | 6·61 | 6·50 | 6·40 | 6·29 | 6·18 | 6·06 | 5·95 |
| 9 | 13·61 | 10·11 | 8·72 | 7·96 | 7·47 | 7·13 | 6·88 | 6·69 | 6·54 | 6·42 | 6·23 | 6·03 | 5·83 | 5·73 | 5·62 | 5·52 | 5·41 | 5·30 | 5·19 |
| 10 | 12·83 | 9·43 | 8·08 | 7·34 | 6·87 | 6·54 | 6·30 | 6·12 | 5·97 | 5·85 | 5·66 | 5·47 | 5·27 | 5·17 | 5·07 | 4·97 | 4·86 | 4·75 | 4·64 |
| 11 | 12·23 | 8·91 | 7·60 | 6·88 | 6·42 | 6·10 | 5·86 | 5·68 | 5·54 | 5·42 | 5·24 | 5·05 | 4·86 | 4·76 | 4·65 | 4·55 | 4·44 | 4·34 | 4·23 |
| 12 | 11·75 | 8·51 | 7·23 | 6·52 | 6·07 | 5·76 | 5·52 | 5·35 | 5·20 | 5·09 | 4·91 | 4·72 | 4·53 | 4·43 | 4·33 | 4·23 | 4·12 | 4·01 | 3·90 |
| 13 | 11·37 | 8·19 | 6·93 | 6·23 | 5·79 | 5·48 | 5·25 | 5·08 | 4·94 | 4·82 | 4·64 | 4·46 | 4·27 | 4·17 | 4·07 | 3·97 | 3·87 | 3·76 | 3·65 |
| 14 | 11·06 | 7·92 | 6·68 | 6·00 | 5·56 | 5·26 | 5·03 | 4·86 | 4·72 | 4·60 | 4·43 | 4·25 | 4·06 | 3·96 | 3·86 | 3·76 | 3·66 | 3·55 | 3·44 |
| 15 | 10·80 | 7·70 | 6·48 | 5·80 | 5·37 | 5·07 | 4·85 | 4·67 | 4·54 | 4·42 | 4·25 | 4·07 | 3·88 | 3·79 | 3·69 | 3·58 | 3·48 | 3·37 | 3·26 |
| 16 | 10·58 | 7·51 | 6·30 | 5·64 | 5·21 | 4·91 | 4·69 | 4·52 | 4·38 | 4·27 | 4·10 | 3·92 | 3·73 | 3·64 | 3·54 | 3·44 | 3·33 | 3·22 | 3·11 |
| 17 | 10·38 | 7·35 | 6·16 | 5·50 | 5·07 | 4·78 | 4·56 | 4·39 | 4·25 | 4·14 | 3·97 | 3·79 | 3·61 | 3·51 | 3·41 | 3·31 | 3·21 | 3·10 | 2·98 |
| 18 | 10·22 | 7·21 | 6·03 | 5·37 | 4·96 | 4·66 | 4·44 | 4·28 | 4·14 | 4·03 | 3·86 | 3·68 | 3·50 | 3·40 | 3·30 | 3·20 | 3·10 | 2·99 | 2·87 |
| 19 | 10·07 | 7·09 | 5·92 | 5·27 | 4·85 | 4·56 | 4·34 | 4·18 | 4·04 | 3·93 | 3·76 | 3·59 | 3·40 | 3·31 | 3·21 | 3·11 | 3·00 | 2·89 | 2·78 |
| 20 | 9·94 | 6·99 | 5·82 | 5·17 | 4·76 | 4·47 | 4·26 | 4·09 | 3·96 | 3·85 | 3·68 | 3·50 | 3·32 | 3·22 | 3·12 | 3·02 | 2·92 | 2·81 | 2·69 |
| 21 | 9·83 | 6·89 | 5·73 | 5·09 | 4·68 | 4·39 | 4·18 | 4·01 | 3·88 | 3·77 | 3·60 | 3·43 | 3·24 | 3·15 | 3·05 | 2·95 | 2·84 | 2·73 | 2·61 |
| 22 | 9·73 | 6·81 | 5·65 | 5·02 | 4·61 | 4·32 | 4·11 | 3·94 | 3·81 | 3·70 | 3·54 | 3·36 | 3·18 | 3·08 | 2·98 | 2·88 | 2·77 | 2·66 | 2·55 |
| 23 | 9·63 | 6·73 | 5·58 | 4·95 | 4·54 | 4·26 | 4·05 | 3·88 | 3·75 | 3·64 | 3·47 | 3·30 | 3·12 | 3·02 | 2·92 | 2·82 | 2·71 | 2·60 | 2·48 |
| 24 | 9·55 | 6·66 | 5·52 | 4·89 | 4·49 | 4·20 | 3·99 | 3·83 | 3·69 | 3·59 | 3·42 | 3·25 | 3·06 | 2·97 | 2·87 | 2·77 | 2·66 | 2·55 | 2·43 |
| 25 | 9·48 | 6·60 | 5·46 | 4·84 | 4·43 | 4·15 | 3·94 | 3·78 | 3·64 | 3·54 | 3·37 | 3·20 | 3·01 | 2·92 | 2·82 | 2·72 | 2·61 | 2·50 | 2·38 |
| 26 | 9·41 | 6·54 | 5·41 | 4·79 | 4·38 | 4·10 | 3·89 | 3·73 | 3·60 | 3·49 | 3·33 | 3·15 | 2·97 | 2·87 | 2·77 | 2·67 | 2·56 | 2·45 | 2·33 |
| 27 | 9·34 | 6·49 | 5·36 | 4·74 | 4·34 | 4·06 | 3·85 | 3·69 | 3·56 | 3·45 | 3·28 | 3·11 | 2·93 | 2·83 | 2·73 | 2·63 | 2·52 | 2·41 | 2·29 |
| 28 | 9·28 | 6·44 | 5·32 | 4·70 | 4·30 | 4·02 | 3·81 | 3·65 | 3·52 | 3·41 | 3·25 | 3·07 | 2·89 | 2·79 | 2·69 | 2·59 | 2·48 | 2·37 | 2·25 |
| 29 | 9·23 | 6·40 | 5·28 | 4·66 | 4·26 | 3·98 | 3·77 | 3·61 | 3·48 | 3·38 | 3·21 | 3·04 | 2·86 | 2·76 | 2·66 | 2·56 | 2·45 | 2·33 | 2·21 |
| 30 | 9·18 | 6·35 | 5·24 | 4·62 | 4·23 | 3·95 | 3·74 | 3·58 | 3·45 | 3·34 | 3·18 | 3·01 | 2·82 | 2·73 | 2·63 | 2·52 | 2·42 | 2·30 | 2·18 |
| 40 | 8·83 | 6·07 | 4·98 | 4·37 | 3·99 | 3·71 | 3·51 | 3·35 | 3·22 | 3·12 | 2·95 | 2·78 | 2·60 | 2·50 | 2·40 | 2·30 | 2·18 | 2·06 | 1·93 |
| 60 | 8·49 | 5·79 | 4·73 | 4·14 | 3·76 | 3·49 | 3·29 | 3·13 | 3·01 | 2·90 | 2·74 | 2·57 | 2·39 | 2·29 | 2·19 | 2·08 | 1·96 | 1·83 | 1·69 |
| 120 | 8·18 | 5·54 | 4·50 | 3·92 | 3·55 | 3·28 | 3·09 | 2·93 | 2·81 | 2·71 | 2·54 | 2·37 | 2·19 | 2·09 | 1·98 | 1·87 | 1·75 | 1·61 | 1·43 |
| $\infty$ | 7·88 | 5·30 | 4·28 | 3·72 | 3·35 | 3·09 | 2·90 | 2·74 | 2·62 | 2·52 | 2·36 | 2·19 | 2·00 | 1·90 | 1·79 | 1·67 | 1·53 | 1·36 | 1·00 |

## Table 3. Percentiles of the chi-square distribution

| d.f. | $X^2_1$ | $X^2_2$ | $X^2_5$ | $X^2_{10}$ | $X^2_{20}$ | $X^2_{30}$ | $X^2_{50}$ | $X^2_{70}$ | $X^2_{80}$ | $X^2_{90}$ | $X^2_{95}$ | $X^2_{98}$ | $X^2_{99}$ | $X^2_{99.9}$ |
|---|---|---|---|---|---|---|---|---|---|---|---|---|---|---|
| 1 | .00016 | .00963 | .0039 | .016 | .064 | .15 | .46 | 1.07 | 1.64 | 2.71 | 3.84 | 5.41 | 6.64 | 10.83 |
| 2 | .02 | .04 | .10 | .21 | .45 | .71 | 1.39 | 2.41 | 3.22 | 4.60 | 5.99 | 7.82 | 9.21 | 13.82 |
| 3 | .12 | .18 | .35 | .58 | 1.00 | 1.42 | 2.37 | 3.66 | 4.64 | 6.25 | 7.82 | 9.81 | 11.34 | 16.27 |
| 4 | .30 | .43 | .71 | 1.06 | 1.65 | 2.20 | 3.36 | 4.88 | 5.99 | 7.78 | 9.49 | 11.67 | 13.28 | 18.46 |
| 5 | .55 | .75 | 1.14 | 1.61 | 2.34 | 3.00 | 4.35 | 6.06 | 7.29 | 9.24 | 11.07 | 13.39 | 15.09 | 20.52 |
| 6 | .87 | 1.13 | 1.64 | 2.20 | 3.07 | 3.83 | 5.35 | 7.23 | 8.56 | 10.64 | 12.59 | 15.03 | 16.81 | 22.46 |
| 7 | 1.24 | 1.56 | 2.17 | 2.83 | 3.82 | 4.67 | 6.35 | 8.38 | 9.80 | 12.02 | 14.07 | 16.62 | 18.48 | 24.32 |
| 8 | 1.65 | 2.03 | 2.73 | 3.49 | 4.59 | 5.53 | 7.34 | 9.52 | 11.03 | 13.36 | 15.51 | 18.17 | 20.09 | 26.12 |
| 9 | 2.09 | 2.53 | 3.32 | 4.17 | 5.38 | 6.39 | 8.34 | 10.66 | 12.24 | 14.68 | 16.92 | 19.68 | 21.67 | 27.88 |
| 10 | 2.56 | 3.06 | 3.94 | 4.86 | 6.18 | 7.27 | 9.34 | 11.78 | 13.44 | 15.99 | 18.31 | 21.16 | 23.21 | 29.59 |
| 11 | 3.05 | 3.61 | 4.58 | 5.58 | 6.99 | 8.15 | 10.34 | 12.90 | 14.63 | 17.28 | 19.68 | 22.62 | 24.72 | 31.26 |
| 12 | 3.57 | 4.18 | 5.23 | 6.30 | 7.81 | 9.03 | 11.34 | 14.01 | 15.81 | 18.55 | 21.03 | 24.05 | 26.22 | 32.91 |
| 13 | 4.11 | 4.76 | 5.89 | 7.04 | 8.63 | 9.93 | 12.34 | 15.12 | 16.98 | 19.81 | 22.36 | 25.47 | 27.69 | 34.53 |
| 14 | 4.66 | 5.37 | 6.57 | 7.79 | 9.47 | 10.82 | 13.34 | 16.22 | 18.15 | 21.06 | 23.68 | 26.87 | 29.14 | 36.12 |
| 15 | 5.23 | 5.98 | 7.26 | 8.55 | 10.31 | 11.72 | 14.34 | 17.32 | 19.31 | 22.31 | 25.00 | 28.26 | 30.58 | 37.70 |
| 16 | 5.81 | 6.61 | 7.96 | 9.31 | 11.15 | 12.62 | 15.34 | 18.42 | 20.46 | 23.54 | 26.30 | 29.63 | 32.00 | 39.29 |
| 17 | 6.41 | 7.26 | 8.67 | 10.08 | 12.00 | 13.53 | 16.34 | 19.51 | 21.62 | 24.77 | 27.59 | 31.00 | 33.41 | 40.75 |
| 18 | 7.02 | 7.91 | 9.39 | 10.86 | 12.86 | 14.44 | 17.34 | 20.60 | 22.76 | 25.99 | 28.87 | 32.35 | 34.80 | 42.31 |
| 19 | 7.63 | 8.57 | 10.12 | 11.65 | 13.72 | 15.35 | 18.34 | 21.69 | 23.90 | 27.20 | 30.14 | 33.69 | 36.19 | 43.82 |
| 20 | 8.26 | 9.24 | 10.85 | 12.44 | 14.58 | 16.27 | 19.34 | 22.78 | 25.04 | 28.41 | 31.41 | 35.02 | 37.57 | 45.32 |
| 21 | 8.90 | 9.92 | 11.59 | 13.24 | 15.44 | 17.18 | 20.34 | 23.86 | 26.17 | 29.62 | 32.67 | 36.34 | 38.93 | 46.80 |
| 22 | 9.54 | 10.60 | 12.34 | 14.04 | 16.31 | 18.10 | 21.24 | 24.94 | 27.30 | 30.81 | 33.92 | 37.66 | 40.29 | 48.27 |
| 23 | 10.20 | 11.29 | 13.09 | 14.85 | 17.19 | 19.02 | 22.34 | 26.02 | 28.43 | 32.01 | 35.17 | 38.97 | 41.64 | 49.73 |
| 24 | 10.86 | 11.99 | 13.85 | 15.66 | 18.06 | 19.94 | 23.34 | 27.10 | 29.55 | 33.20 | 36.42 | 40.27 | 42.98 | 51.18 |
| 25 | 11.52 | 12.70 | 14.61 | 16.47 | 18.94 | 20.87 | 24.34 | 28.17 | 30.68 | 34.38 | 37.65 | 41.57 | 44.31 | 52.62 |
| 26 | 12.20 | 13.41 | 15.38 | 17.29 | 19.82 | 21.79 | 25.34 | 29.25 | 31.80 | 35.56 | 38.88 | 42.86 | 45.64 | 54.05 |
| 27 | 12.88 | 14.12 | 16.15 | 18.11 | 20.70 | 22.72 | 26.34 | 30.32 | 32.91 | 36.74 | 40.11 | 44.14 | 46.96 | 55.48 |
| 28 | 13.56 | 14.85 | 16.93 | 18.94 | 21.59 | 23.65 | 27.34 | 31.39 | 34.03 | 37.92 | 41.34 | 45.42 | 48.28 | 56.89 |
| 29 | 14.26 | 15.57 | 17.71 | 19.77 | 22.48 | 24.58 | 28.34 | 32.46 | 35.14 | 39.09 | 42.56 | 46.69 | 49.59 | 58.30 |
| 30 | 14.95 | 16.31 | 18.49 | 20.60 | 23.36 | 25.51 | 29.34 | 33.53 | 36.25 | 40.26 | 43.77 | 47.96 | 50.89 | 59.70 |

Adapted from Table IV of Fisher and Yates: Statistical Tables for Biological, Agricultural and Medical Research, published by Longman Group Ltd., London (previously published by Oliver and Boyd, Edinburgh), with permission.

# Glossary of Symbols and Abbreviations

In the following glossary of symbols and abbreviations that are used throughout this book, Roman letters appear first, Greek letters next, and miscellaneous symbols at the end. Symbols that occur only in Chapter 11, or the meaning of a symbol that is specific to Chapter 11, are not included. Alternative symbols that are also often used, and which occur in some of the problems, are indicated in parentheses.

| | |
|---|---|
| AR | attributable risk |
| $b_0$ | sample intercept |
| $b_1, b_2, \ldots$ | sample regression coefficients |
| d | deviation from the sample mean; difference between paired values |
| d.f. | degrees of freedom |
| CV | coefficient of variation |
| e | irrational mathematical constant equal to about 2.71828; expected number of counts in a table (also denoted E); estimated error or residual from a regression model |
| F | percentile of Snedecor's F distribution or corresponding test statistic |
| $f(y)$ | (probability) density function |
| $F(y)$ | cumulative (probability) distribution function |
| $g_2$ | fourth cumulant; the coefficient of kurtosis minus three |
| log | logarithm |
| $\log_e$ | logarithm to base e; natural logarithm |
| Hg | mercury |
| $H_0$ | null hypothesis |
| I | incidence |
| l | liter |

| | |
|---|---|
| mg | milligram |
| ml | milliliter |
| MS | mean square |
| n | sample size; parameter of the binomial distribution |
| o | observed number of counts in a table (also denoted O) |
| OR | odds ratio |
| p | sample proportion (also used to denote the probability parameter of the binomial distribution); p-value (also denoted P) |
| $P(A)$ | probability of the event A |
| $P(A \mid B)$ | conditional probability of the event A given B |
| r | correlation coefficient |
| R | multiple correlation coefficient |
| $R^2$ | proportion of the variability explained by a regression model |
| s | sample standard deviation (estimate) |
| $s^2$ | sample variance (estimate) |
| $s_{xy}$ | sample covariance of X and Y (estimate) |
| S | sample standard deviation (estimator) |
| $S^2$ | sample variance (estimator) |
| s.e.m. | standard error of the mean |
| SS | sum of squares |
| t | percentile of Student's t distribution or corresponding test statistic |
| T | rank sum statistic |
| $x^2$ | sample chi-square statistic (also denoted $X^2$) |
| x, y | particular values of (random) variables |
| X, Y | random variables |
| $\bar{y}$ | sample mean of y (estimate) |
| $\bar{Y}$ | sample mean of Y (estimator) |
| $\hat{y}$ | value of y predicted from a regression equation |
| z | particular value of a standardized random variable |
| Z | standardized random variable |
| | |
| $\alpha$ | probability of Type I error; significance level |
| $\beta$ | probability of Type II error; complement of power |
| $\beta_0$ | population intercept |
| $\beta_1, \beta_2, \ldots$ | population regression coefficients |

| | |
|---|---|
| $\epsilon$ | error or residual from a regression model |
| $\lambda$ | parameter of the Poisson distribution |
| $\mu$ | population mean |
| $\pi$ | irrational mathematical constant equal to the ratio of the circumference of a circle to its diameter; population proportion |
| $\sigma$ | population standard deviation |
| $\sigma^2$ | population variance |
| $\Sigma$ | summation |
| $\chi^2$ | percentile of the chi-square distribution (also used to denote the corresponding statistic) |
| $=$ | equals |
| $\doteq$ | equals approximately |
| $\pm$ | plus or minus |
| $<$ | is less than |
| $\leq$ | is less than or equal to |
| $>$ | is greater than |
| $\geq$ | is greater than or equal to |
| $/$ | divided by, per |
| $\%$ | percent |
| $\sqrt{\phantom{x}}$ | square root |
| $!$ | factorial |
| $\infty$ | infinity |

# Index

A capital "F" following a page number indicates a figure; a capital "T" indicates a table.

graphs in
bar-graph, 36
construction of, guidelines for, 34–35
cumulative plot, 37–38, 39F, 40
frequency polygon, 36–37, 37F, 38F
histogram, 35–36, 35F, 36F
scatter plot, 40, 40F
tree diagram, 40–41, 41F, 74F, 76F
measures of central tendency in, 47–49
measures of shape in, 52–53, 53F
measures of spread or variability in, 49–52
need for, 30–31
practice problems in, 55–59
predictive values in, 47
proportions in, 41–44
rates in, 41–44
scales of measurement and, 31–32
sensitivity values in, 46
specificity values in, 46
tables of, 32–34, 33T
Deviation(s). *See also* Standard deviation
regression and, 186–187, 187F
variance computation and, 50
vertical and horizontal, 191
d.f. *See* Degrees of freedom
Dichotomous trait, binomial distribution and, 85–88, 86F
Difference between two means, confidence limits for, 119–121
Discrepant data, routine variability vs., 3
Discrete data, 31
Discriminant analysis, 235–236
Disease
causes of, observational studies of, 16–18
incidence of, 43–44
prevalence of, 43–44
study of, cause and effect in, 10–11
Disease frequency, relative measures of, 44–46
Disjunction, logical, 129
Distribution(s)
chi-square, 155–157, 157F,

276–277. *See also* Chi-square distribution
F, 137–138
analysis of variance and, 211, 212–213. *See also* Analysis of variance
multiple regression and, 198
simple linear regression and, 188
Student's t test and, 279
tables of values for, 285–289
normal, standard, 111–114, 111T, 112F, 113F
Student's t vs., 117, 117F
probability, 98–99
binomial, 85–89, 86F, 94F. *See also* Binomial distribution
definition of, 85
Gaussian, 94–96, 94F
normal, 94–96, 94F
Poisson, 90–92, 92F
practice problems in, 100–102
rectangular, 92–94, 93F
uniform, 92–94, 93F
"rare events," 90–92, 92F
Student's t, 116–118, 117F
F-test and, 279
population mean hypothesis testing and, 128–131, 129F
table of percentiles for, 284
Distribution-free methods, 141
Distribution functions, cumulative, 96–98, 97F
Double blinding, 24
Duncan's procedure, 280

EFFECT(S)
additive, 220, 220F
fixed, 215
interaction, 220–221, 221F
main, 219
random, 215
simple, 218
Efficient estimator, 107
Epidemiological studies, observational, 16
ε. *See* Error
Equality
means, test of, 138–140
medians, test of, 140–144
variances, test of, 136–138
Error
estimated, 186

Split-plot design, 21
Spread, measures of, 49–52
S-shaped, curve, 237, 237F
Standard deviation, 51
  estimator, 104
  notation for, 105
Standard error, 104–105
Standard normal distribution,
      111–114, 111T, 112F, 113F
  Student's t vs., 117, 117F
Standardized random variable, 111
Statistic
  definition of, 3
  parameter vs., 13
Statistical inference, *See also*
      Inference
  specific examples of, 6–7
Statistical methods, appropriateness
      of, 249–250
Statistician(s)
  contributions of, acknowledgement
      of, 6
  research process and, 5
  research process needs of, 5–6
Statistics. *See also* Data
  definition of, 3
  descriptive, 53–55
    appropriate use of, 248–249
    disease frequency in, 44–46
    graphs in
      bar-graph; 36
      construction of, guidelines
        for; 34–35
      cumulative plot, 37–38, 39F,
        40
      frequency polygon, 36–37,
        37F, 38F
      histogram, 35–36, 35F, 36F
      scatter plot, 40, 40F
      tree diagrams, 40–41, 74F,
        76F
    measures of central tendency in,
      47–49
    measures of shape in, 52–53,
      53F
    measures of spread or variability
      in, 49–52
    need for, 30–31
    practice problems in, 55–59
    predictive values in, 47
    proportions in, 41–44
    rates in, 41–44
    scales of measurement and,

31–32
  sensitivity values in, 46
  specificity values in, 46
  tables of, 32–34, 33T
relevance of, 1–8
role of, 1–8
Stepwise regression
  backward, 200
  forward, 199
Straight line, 183–184, 183F, 184F.
      *See also* Regression
Strata, 14
Stratified random sample, 14
Streptomycin, tuberculosis and,
      classic study of, 7
Student's t distribution, 116–118,
      117F
  F test and, 279
  population mean hypothesis testing
      and, 128–129, 129F
  table of percentiles for, 284
Study design(s), 5. *See also* Research
      project(s)
  change-over, 21
  completely randomized, 20
  confounding and, 10
  critical evaluation of, 247
  cross-over, 21
  experimental, 19
  factorial arrangement in, 22
  mistakes in, 6
  randomized, single classification
      and, analysis of variance
      in, 212–214, 212T, 213T
  randomized blocks, 20–21
  sequential, 21–22
  split-plot, 21
Study population. *See also*
      Population(s), 13
  information collection about, 13
  target population vs., 13
Study unit(s). *See also* Sample(s)
  control and, 21
  definition of, 12–13
  experimental, 18
    randomization and, 19–20
  multiple, multiple measures vs.,
      13
  number of, experimental studies
      and, 23
  number of, number of measures
      on vs., 12–13
  selection of